THE HUMAN DAWN

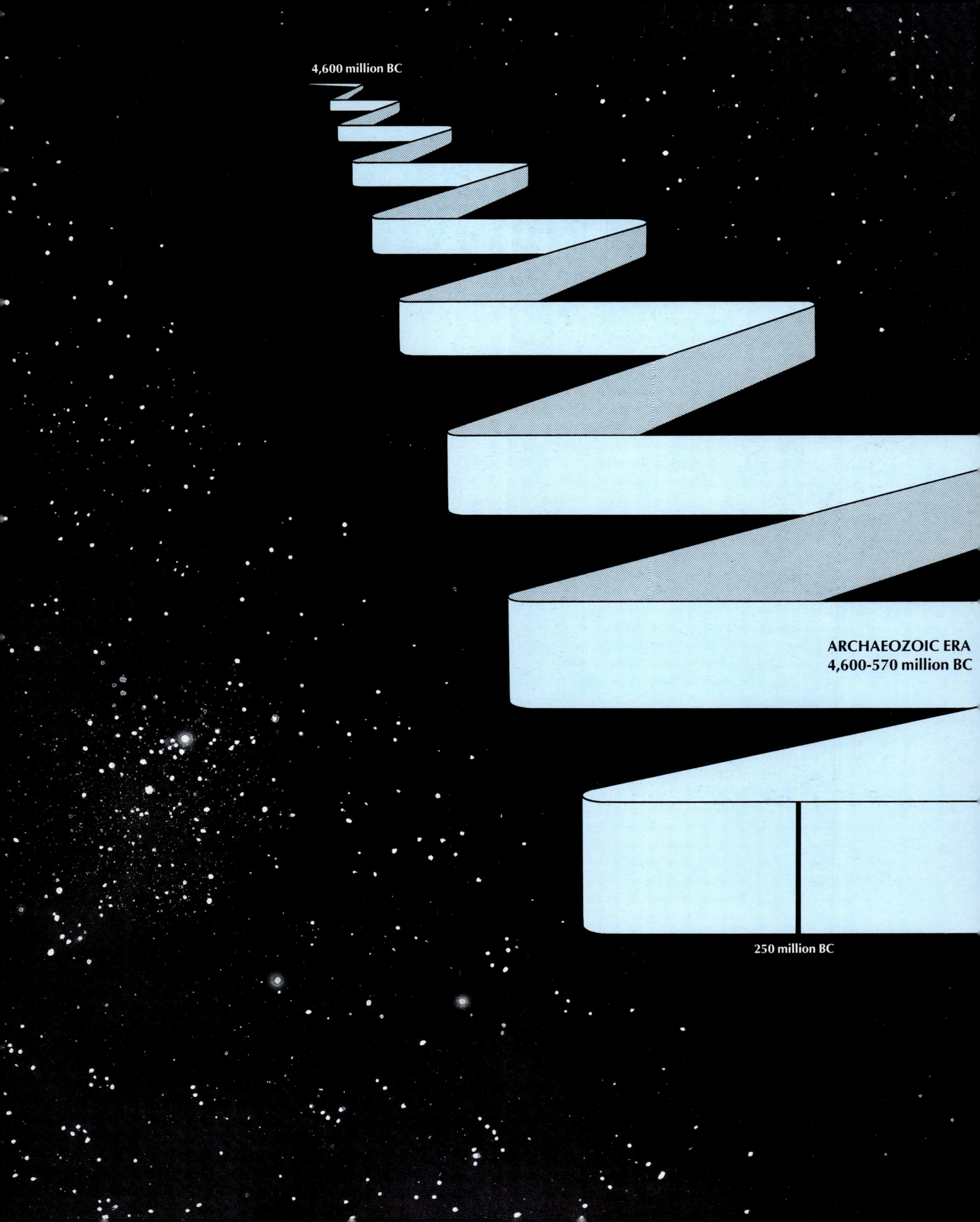
4,600 million BC
ARCHAEOZOIC ERA
4,600-570 million BC
250 million BC

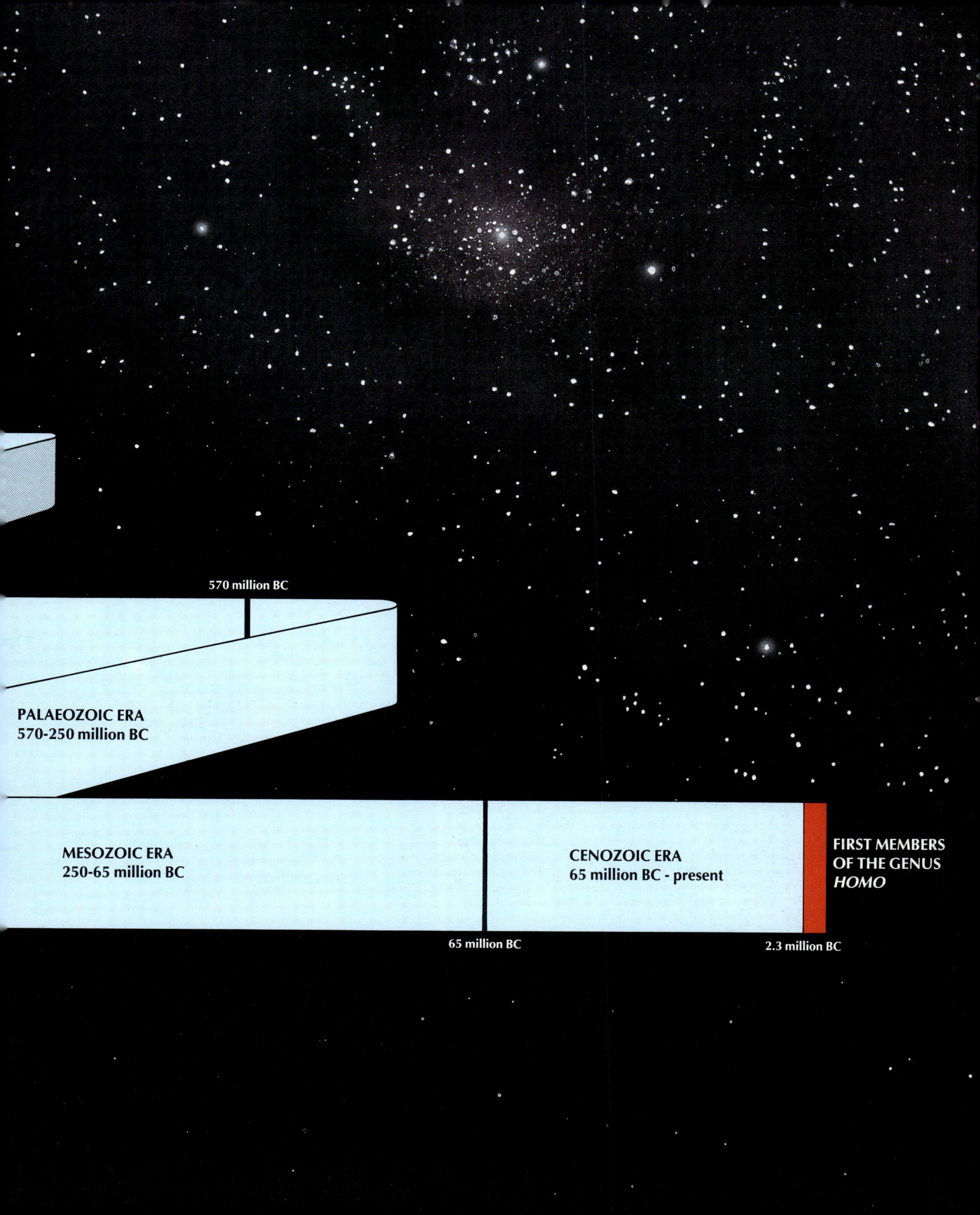
570 million BC
PALAEOZOIC ERA
570-250 million BC
MESOZOIC ERA
250-65 million BC
CENOZOIC ERA
65 million BC - present
FIRST MEMBERS OF THE GENUS *HOMO*
65 million BC
2.3 million BC

TIME LIFE BOOKS

LIBRARY OF CURIOUS AND UNUSUAL FACTS
BUILDING BLOCKS
A CHILD'S FIRST LIBRARY OF LEARNING
VOYAGE THROUGH THE UNIVERSE
THE THIRD REICH
MYSTERIES OF THE UNKNOWN
TIME-LIFE HISTORY OF THE WORLD
FITNESS, HEALTH & NUTRITION
HEALTHY HOME COOKING
UNDERSTANDING COMPUTERS
THE ENCHANTED WORLD
LIBRARY OF NATIONS
HOME REPAIR AND IMPROVEMENT
PLANET EARTH
THE SEAFARERS
CLASSICS OF WORLD WAR II
THE GOOD COOK
THE WORLD'S WILD PLACES
TIME-LIFE LIBRARY OF ART

This volume is one in the Time-Life series
HISTORY OF THE WORLD.

THE HUMAN DAWN

BY THE EDITORS OF TIME-LIFE BOOKS

TIME-LIFE BOOKS, AMSTERDAM

TIME-LIFE BOOKS

EUROPEAN EDITOR: Ellen Phillips
Design Director: Ed Skyner
Director of Editorial Resources: Samantha Hill
Chief Sub-Editor: Ilse Gray
Assistant Design Director: Mary Staples

TIME-LIFE HISTORY OF THE WORLD

SERIES EDITOR: Charles Boyle

Editorial Staff for *The Human Dawn*
Editor: Chris Middleton
Designer: Lynne Brown
Researchers: Marie-Louise Collard, Tim Fraser, Louise Tucker
Sub-Editors: Luci Collings, Tim Cooke
Design Assistant: Rachel Gibson
Editorial Assistant: Molly Sutherland

Picture Department
Picture Administrator: Amanda Hindley
Picture Coordinator: Elizabeth Turner

Editorial Production
Production Assistant: Emma Veys
Editorial Department: Theresa John, Debra Lelliott, Juliet Lloyd-Price

ISBN 0 7054 0989 9

Volume Consultant:

CHRIS SCARRE, Editor of the *Cambridge Archaeological Journal*, Macdonald Institute for Archaeological Research, Cambridge.

SPECIAL CONTRIBUTORS: Stephen Downes, Ellen Galford, Roy Hayward, Michael Kerrigan, Alan Lothian (text); Judy Aspinall, Barbara Hicks, Rebecca Hunter (research).

Correspondents: Elisabeth Kraemer-Singh (Bonn); Christina Lieberman (New York); Maria Vincenza Aloisi (Paris); Ann Natanson (Rome). Valuable assistance was also provided by: Sasha Isachenko (Moscow), Josephine du Brusle (Paris), Traudl Lessing (Vienna).

CONTENTS

PROLOGUE TO HUMANKIND

Most of our knowledge about life before people is derived from fossils – traces of ancient life embedded in rock. The fossilization process began when animals and plants died and became buried in sediment such as mud or sand. The water and bacteria in the sediment then slowly dissolved the dead organism, while at the same time impregnating the hard parts with mineral deposits. Eventually a shape formed that was an exact replica, except that it was made of mineral rather than organic matter. Over millions of years, the sediment drained of water and was compacted – together with the fossilized shape inside it – into solid rock.

Of the creatures that lived during the first 4,000 million years of the earth's existence *(chronology, below)*, most left no trace; their remains were not made of a durable enough material. However, around 500 million BC there emerged creatures with hard skeletal structures that did not immediately decompose after death. These included the vertebrates – animals with a backbone. Their fossil record begins with traces of bony fishes which had neither fins nor jaws; they probably wriggled through the water like tadpoles and fed by sucking in plankton. Later species developed teeth, hinged jaws and fins, the most spectacular example being the shark, which appeared around 375 million BC.

From bony, finned fishes developed amphibians, the first vertebrates to walk on land. Like fish, these creatures laid their eggs in the water, but unlike fish they could live on land as well. Pioneers of this life style are thought to have been the *Crossopterygii*, fish possessed of muscular fins with which they dragged themselves across land, and lungs which were able to take in air direct.

Over the course of time, amphibians developed legs and feet that operated more efficiently on land than did fins. Some smaller species also began to spend less time in and around water, having been driven inland by larger rivals, and they developed characteristics that reflected this drier environment. They grew harder, scalier skins, and laid eggs that had shell coatings, to protect the embryo inside from desiccation. These were the very first reptiles. The fossil shown on the right is that of *Seymouria baylorensis*, an 80-centimetre-long creature that had a strong reptilian backbone fused to the pelvis, while at the same time possessing distinct amphibian characteristics, including labyrinthine teeth and, located at either side of the back of the skull, a notch into which the eardrums fitted.

Reptiles did not stay small: *Brachiosaurus*, the biggest of the dinosaurs, measured 25 metres from head to tail, and weighed 80 tonnes. But, around 65 million BC, the dinosaurs died out, for mysterious reasons – possibly the collision of a giant meteor with the earth, possibly because the reptiles, being cold-blooded and dependent on the sun for energy, were vulnerable to any sudden drops in temperature. Their mastery of the earth passed down to the mammals, which were much smaller, warm-blooded creatures that suckled their young (*mamma* is the Latin word for breast) and had evolved a variety of internal mechanisms for maintaining a steady blood temperature, which

ARCHAEOZOIC ERA
(4,600-570 million BC)

PRECAMBRIAN LIFE
- Formation of the earth (4,600 million BC)
- First life (algal limestone pillars) (3,300 million BC)
- First animals (sea worms, jellyfish) (680 million BC)

PALAEOZOIC ERA
(570-250 million BC)

CAMBRIAN (570-510 million BC)
- Invertebrates evolved (sponges, molluscs, shellfish)

ORDOVICIAN (510-435 million BC)
- First vertebrates (500 million BC)
- First fish (470 million BC)

SILURIAN (435-410 million BC)
- First plants on land (412 million BC)

DEVONIAN (410-355 million BC)
- First insects (380 million BC)
- First amphibians (360 million BC)

CARBONIFEROUS (355-290 million BC)
- First tropical forests (330 million BC)
- First reptiles (300 million BC)

PERMIAN (290-250 million BC)

MESOZOIC ERA
(250-65 million BC)

TRIASSIC (250-205 million BC)
- Conifers and ferns proliferated
- First dinosaurs (215 million BC)

JURASSIC (205-140 million BC)
- Continents split
- First mammals (190 million BC)
- First birds (140 million BC)

CRETACEOUS (140-65 million BC)
- Many species died out, including dinosaurs (65 million BC)

meant that they relied not on the sun but on food for their energy.

From these early, mouse-sized creatures evolved larger mammals, among them a group of tree-dwelling animals with long tails, grasping paws and eyes at the front rather than at the sides of the head. This last-named feature greatly improved the judgment of distance, essential for an animal moving quickly from branch to branch. These were the primates, from which, more than 20 million years ago, evolved the first apes.

The journey from bony fish to apes spanned three evolutionary periods: the Palaeozoic, the Mesozoic and the Cenozoic, the eras of old, middle and new life respectively *(chronology, below)*. The fossils of some of the creatures that lived during those eras appear on the following pages. The evolution of their distinctive physical features shows the effects on these animals of the changing configurations of the continents that took place during this period – a phenomenon that was caused by continual movement of the earth's landmasses across the ocean floor.

CENOZOIC ERA
(65 million BC - present)

PALAEOGENE (65-25 million BC)
- Many mammals evolved (65-55 million BC)
- First primates (60 million BC)
- First horses (55 million BC)
- First sea mammals (whales) (54 million BC)

NEOGENE (25-2 million BC)
- First apes (23 million BC)
- First australopithecines (5 million BC)
- First *Homo* (2.3 million BC)

QUATERNARY (2 million BC - present)
- Most animal species in existence
- First civilizations (8000 BC)

FROM FISH TO REPTILE

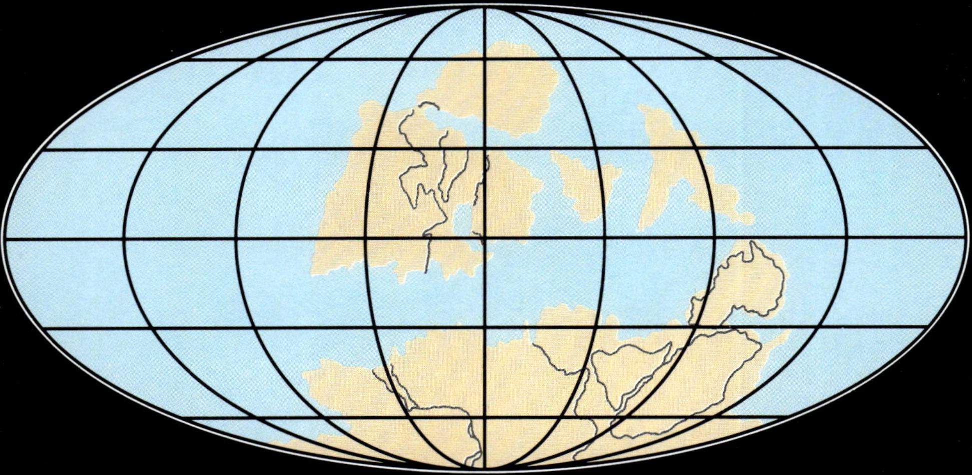

In the middle of the Palaeozoic era, one giant continent – comprising modern-day South America, Africa, India, Australia and Antarctica – sat in the southern hemisphere *(map, left)*, while in the north lay the landmasses that would become North America, Europe and Asia.

In the deep seas around the continents, vertebrate fish thrived. However, around 400 million BC, geological convulsions created new mountain ranges and lakes – habitats that attracted first plants, then insects and fish on to dry land. By the end of the era some amphibians had evolved into exclusively land-based reptiles.

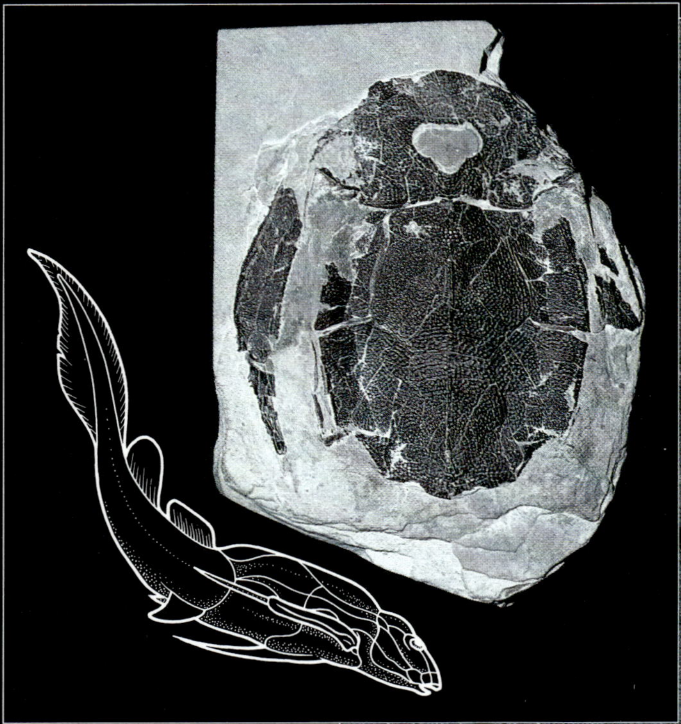

Bony armour protected *Bothriolepis canadensis*, shown above in fossil and reconstructed form, against predators. This 360-million-year-old fish never left the water, where its jaws and fins gave it an added advantage over less developed species.

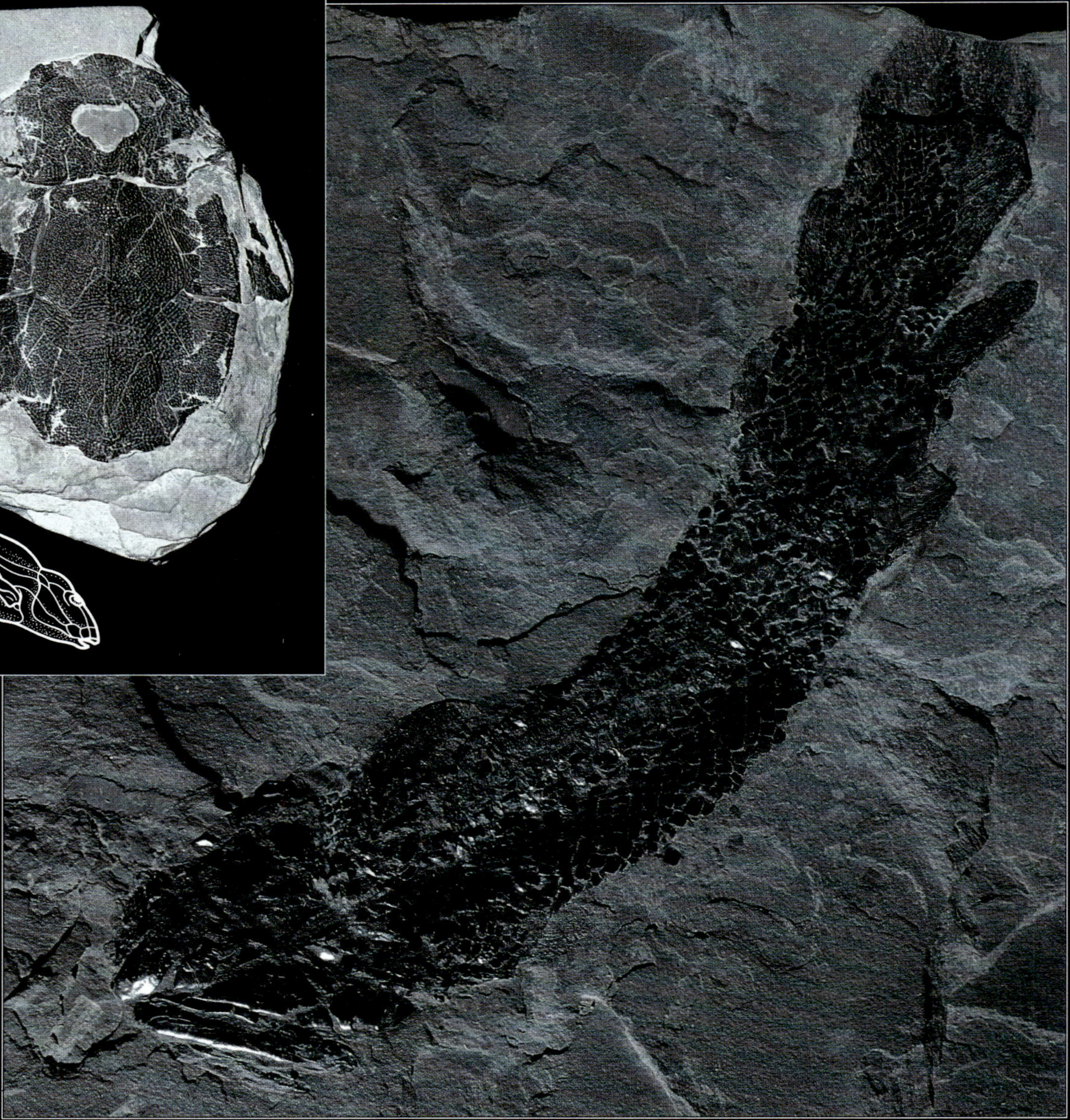

The strong, rounded fins of the 30-centimetre-long *Gyroptychius*, from around 380 million BC, indicate a fish that is evolving into an amphibian. Such fins gradually became limbs.

The fins have become fully formed legs in this fossilized imprint of a completely amphibian *Branchiosaurus*, which lived 300 million years ago. The creature retains its fishlike tail, however, for use in the water.

Measuring little more than half a metre from head to tail, this ***Labidosaurus hamatus*** was one of the world's first reptiles, living entirely on land at the end of the Palaeozoic era.

The two-metre-long *Askeptosaurus italicus*, a reptilian ancestor of the lizard, abandoned its land-based life style and returned to live in the warm seas near the coast of what is now Italy.

Powerful hind legs and a slender build mark our *Coelophysis bauri* as having been a swift and effective hunter. Three metres long, this rapacious creature – one of the earliest known carnivorous dinosaurs – chased many species into extinction around 215 million BC

A rock fragment bears the fossil of 220-million-year-old *Eudimorphodon ranzii*, the earliest known flying reptile. *Eudimorphodon* had three clawed fingers on each hand, and one huge, elongated finger, which in this specimen sprouts out like an extension to the arm. The wings, of membrane-like skin, were attached to this finger, and measured 75 centimetres fully spread. Rather than compete with the dinosaurs for food on land, *Eudimorphodon* probably flew low over water, searching out fish.

THE EVOLUTION OF MAMMALS

By 170 million BC, the southern landmass had joined up with its northern counterparts *(map, right)*, resulting in the creation of one vast continent, with a great gulf lying to the east.

The conditions that prevailed throughout this Pangaea, or "whole world", were warm and rather humid. Lush tropical forests flourished, and in them thrived the giant, leaf-browsing dinosaurs, who with their carnivorous contemporaries came to dominate all other land vertebrates. Some of the smaller reptiles took to the air, others to the sea.

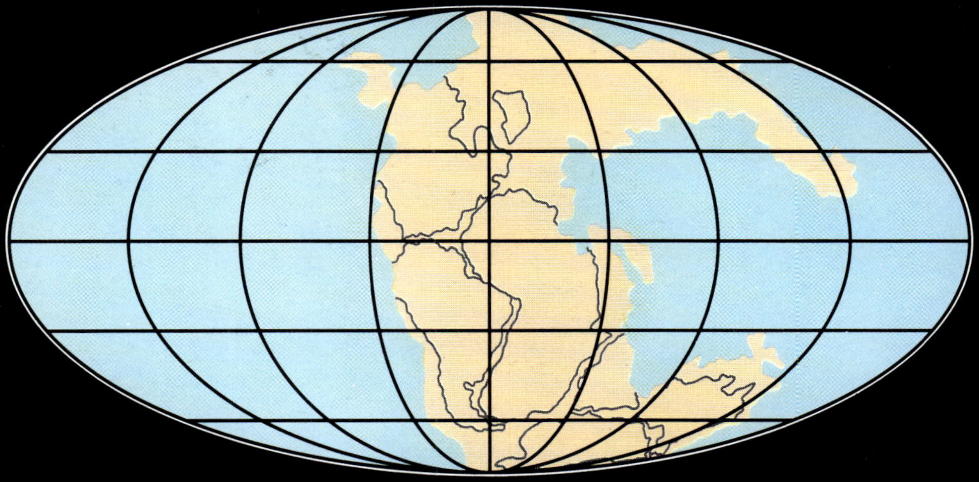

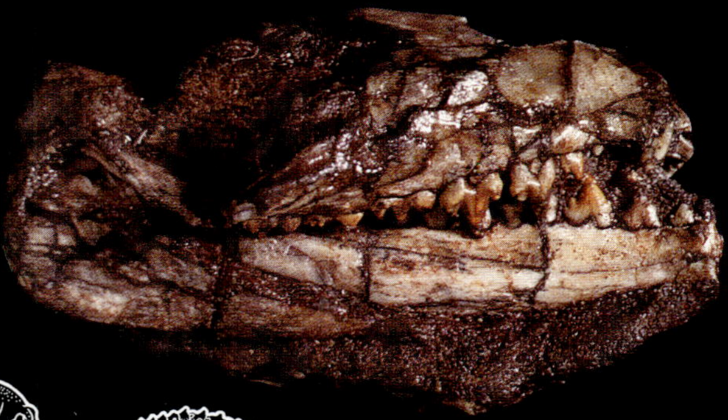

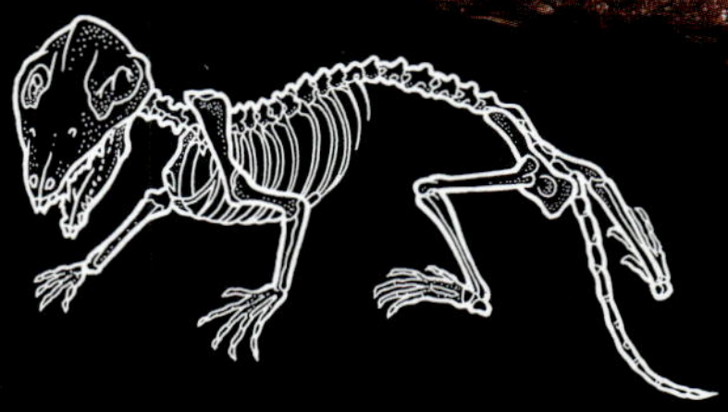

This fossilized head *(above)* and reconstructed skeleton *(left)* belong to *Megazostrodon*, a tiny, shrewlike animal that lived in Africa about 190 million years ago. One of the earliest known mammals, *Megazostrodon* probably hunted for grubs and insects at night, to avoid carnivorous dinosaurs. Like its reptilian ancestors it laid eggs, but unlike them suckled its newly hatched young – an exclusively mammalian characteristic. Even the largest of these creatures grew to only 12 centimetres.

Based on fossilized remains *(left)*, a drawing of a pantothere *(right)* – the direct ancestor of almost all later mammals – shows a 20-centimetre-long creature well adapted for tree life, with a long tail for balance and powerful hind legs with which to launch itself from branch to branch. Like *Megazostrodon*, the pantothere avoided dinosaurs by a policy of nocturnal feeding.

THE APPEARANCE OF APES

By 50 million BC the world was beginning to look as it does today *(map, right)*. Australia was still joined to Antarctica, and India was adrift in mid-ocean, but the Americas had separated from Africa and Europe. This, the Cenozoic era, witnessed the formation of the Alps, the Himalayas, the Rockies and the Andes, as well as the evolution of flora and fauna similar to those with which we are familiar today. A warm climate stimulated the growth of forests, in which mammals freely multiplied now that the threat from dinosaurs had been removed.

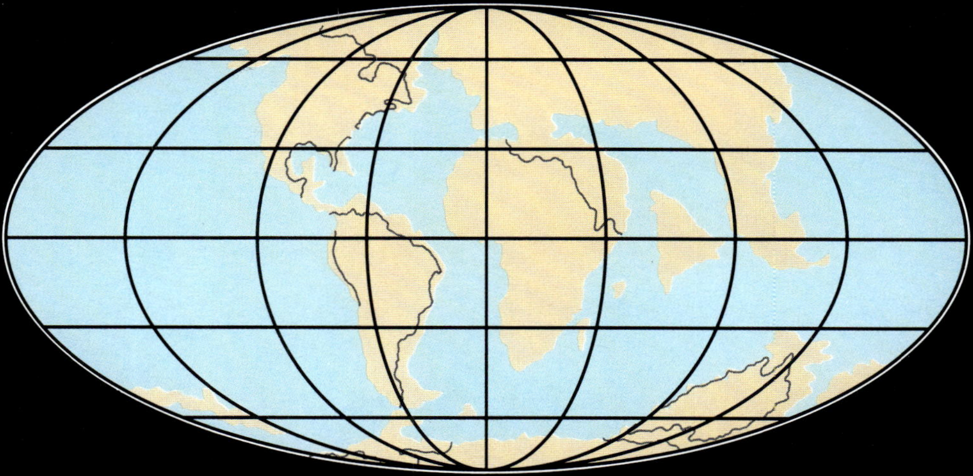

A restored skeleton of 50-million-year-old, lemur-like *Smilodectes (opposite page)* exhibits the characteristic features of a primate: grasping hands, enlarged brain area and a snout that had grown shorter because an acute sense of smell was not essential for this tree-living animal, which fed only on fruit and insects. *Smilodectes'* eyes are positioned at the front of the skull, permitting the visual fields to overlap. These eyes would have transmitted a single picture to the brain, making it easier to judge distances; animals with eyes on the sides of their skull would have seen two images.

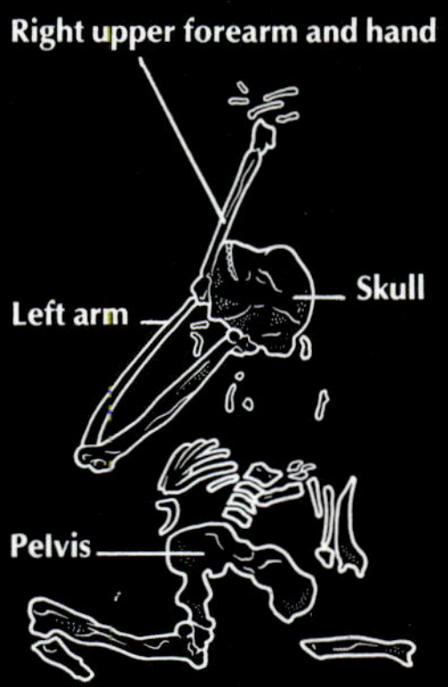

A fossil *(left)* from a Tuscan coal bed bears the imprint of *Oreopithecus*, the mountain ape, which lived at the end of the Cenozoic era, around 14 million BC. A drawing *(above)* based on the fossil shows that although *Oreopithecus* had the long arms of an ape, it also had a sufficiently broad pelvis to sustain a semiupright posture. In this fossil, for the first time, humans are foreshadowed.

THE PATH OF EVOLUTION

1

On a warm afternoon in Africa, 3.5 million years ago, something in the long grass stirred. Antelopes at the lake's edge stopped drinking and looked up, their nostrils sifting the air for scent. They scanned the surrounding countryside, searching first the open savanna, its pale strands shuddering in the gentle breeze, then the darker woodland behind, which stretched far up into the foothills before giving way to bald, parched, mountain slopes. The only movement visible was high above, where a billowing white cloud travelled serenely across the sky, its aerial path mirrored on the ground by a large, slow-moving pool of shadow.

What was about to take place was never witnessed by human eyes, for the simple reason that 3.5 million years ago, human beings still did not exist. The only spectators at this re-created scene were the small clusters of birds and animals that had come down to the lake that afternoon to drink.

Whatever it was in the grass moved again. A group of gazelles nearby, sighting the disturbance, shifted nervously in the lake's muddy shallows. There was silence for a second, then, rising timorously above the line of light brown stalks, there appeared a tiny, dark brown head, in which a pair of eyes darted cautiously from side to side. The features that framed those eyes were those of an ape: flat-topped skull, low forehead, bony ridge over the eyes, nose flattened against the face, and the jaws splayed wide athwart an almost horizontal chin. But there was something decidedly different about this creature, something strangely steady and measured about the way that it moved through the grass, showing no signs of the usual arm-dangling, lop-sided gait that had fitted apes so well for life in the trees, but had always rendered them clumsy and vulnerable when down on the ground.

Slowly yet purposefully, the animal advanced. As it reached the edge of the long grass it paused once more to scan the surrounding countryside for possible predators. Then, with one bold step, it emerged from the undergrowth, not crawling on all fours but walking straight and upright on two feet, its arms not trailing along the ground but gently swinging at its side. Without the cover of the savanna, its diminutive stature became instantly apparent; although a fully developed female of some 20 summers, the creature stood little more than one metre from the top of her hairy head to the tips of her long, slender toes.

To the gazelles and antelopes at the lake's edge, this lone, diminutive figure warranted no more than a passing glance; straight away they assessed her as presenting little threat to them, and returned to their drinking.

But this small creature venturing out for a drink some 3.5 million years ago, in the Afar region of eastern Ethiopia, represents a crucial and pivotal point in the whole epic saga of mankind's evolution from apes to human beings. For she belonged to a group of mammals known as *Australopithecus afarensis*, literally the southern apes

Striding across the Laetoli region of Tanzania, footprints offer incontrovertible evidence that human ancestors walked under the East African sun over 3.5 million years ago. The largest prints, thought to be those of an adult male, were overprinted by steps of a smaller hominid – possibly a female following in his wake – while a younger companion, perhaps a child, walked alongside. At the turn of the savanna season from dry to wet, the three walkers left their tracks in a layer of volcanic ash that rain had turned to a cement-like mud; after they had passed, the sun baked their footprints hard. Later, ash falls covered and preserved these marks of the genus *Australopithecus*, the first hominids to walk fully erect, leaving the hands free.

from Afar, who were the first creatures known to have been fully bipedal – that is to say, to walk on two feet as the rule rather than the exception. What is more, although these australopithecines mostly lived and slept out in the open air, and although they shared a great many of their physical characteristics with gorillas, baboons and chimpanzees, they stood on an evolutionary path that was eventually to lead to modern-day men and women. Thus, although our Ethiopian water-seeker may have been naked and hairy, and may even have got down on all fours to drink from the lake, she can claim direct ancestry with her modern-day counterpart who sits on a bar stool sipping a cocktail.

Indeed, had they but known it, the animals gathered at the lakeside that day were looking at a creature whose descendants would inherit the earth. For although our female australopithecine would be dead within 10 years – few of them lived past 30 – the generations that followed her would develop brains that grew ever larger and hands that grew ever more dexterous. It was this mutually enriching interaction between mental and manual skill that over the next three million years would fuel human beings' genetic journey from ape to anatomically modern man, and would in the process enable them to create an armoury of weapons, tools and tactics that would ensure their eventual domination of every other species on the earth.

The planet on which the australopithecines were struggling to gain a foothold had already been in existence for a great many millions of years. According to the current estimates of astrophysicists, it was approximately 15,000 million years ago that a cosmic fireball of inconceivable magnitude suddenly exploded. Multitudes of galaxies were formed out of the fragments of this explosion, and in one of these, clouds of gas condensed and formed a medium-sized star, our sun. Other gas clouds, circling around this star, condensed too, forming a number of planets around 4,600 million years ago. One of these planets was earth.

The enormous, mind-bending time span involved is, for most people, an obstacle to understanding the events of the remote past. It was so much easier when scholars could simply accept the calculations of James Ussher, an Irish archbishop who, in 1650, added up the number of generations recorded in the Bible and pronounced that the world had been created in the year 4004 BC. This estimate was later refined by Bishop John Lightfoot, vice chancellor of Cambridge University, who further narrowed the hour of humanity's creation down to 9.00 a.m. on October 23, 4004 BC. Nevertheless, it is possible to put the story of evolution into some kind of manageable perspective by adopting a rough timescale that condenses the 4,600 million years of the earth's past into one calendar year.

Working on this basis, the earth's atmosphere would have been unbreatheable for the first seven months, and it was only after 10 months had passed that anything bigger than microscopic organisms started appearing, when the interaction of sunlight and bacteria began releasing oxygen into the atmosphere – the process known as photosynthesis. Fish followed a fortnight later, and seven days after this plants started growing on land. The next nine days saw much activity, with the arrival of insects, amphibians, forests and reptiles, and it was just 15 days ago that the earth first felt the ponderous tread of dinosaurs.

Mammals followed the next day, and for some as yet unknown reason managed to survive when the dinosaurs, possibly due to the earth being devastated by a meteorite, suddenly vanished off the face of the earth a mere five days ago. If we take present time as being 12 noon on a Monday morning, then the first apes arrived on the Friday afternoon before, the australopithecines at 6.00 a.m. this morning, and anatomically modern man, known to scientists as *Homo sapiens sapiens*, only turned up 11 minutes ago. More recently, it has been one minute since the birth of agriculture, 20 seconds since the building of the first city, and only 1.5 seconds since the American Declaration of Independence.

This brief, condensed chronology outlines just the bare bones, so to speak, of what is a complex evolutionary story, but within the chemical and biological constitutions of the creatures that participated in it, there occurred developments of a less visible but no less significant nature. The first of these major events was the creation of life itself, which took place around 3,300 million years ago. On this subject we can say nothing with any degree of certainty, although much can be surmised by tracing genetic changes backwards from life forms that exist today. For example, we know that all living creatures are composed of cells – small, separate units, each of which has the capacity to reproduce itself, thanks to the presence within its confines of a substance commonly known as DNA, or deoxyribonucleic acid. Each cell contains, in the form of a double helix, two intertwined strands of DNA molecules, and it is the quantity and configuration of those molecules that provides the blueprint for the individual cell's genetic characteristics.

In order for a cell to create a replica of itself, it must assemble a new, matching pair of strands. If the assembly is accurate, this new cell will be an exact copy of the original. In this way, cells have the capacity to continue making copies of themselves ad infinitum. However, nature has never been so benevolent as to permit indiscriminate and continual reproduction. Living things might have the right to be born, but nature has always been more sparing when it comes to the right to survive. In prehistoric times, as now, survival was something that a species could only acquire through a gradual process of adapting to its circumstances. And the only long-term

APES AND HUMANS

For countless millenniums, apes and human beings travelled a common evolutionary road until, around five million years ago, human precursors finally stood upright. This demanded the growth of new bone and muscle structures. Later, and more importantly, the brain developed as a control centre for new patterns of movement and behaviour. As illustrated on page 23, the progression from early hominid to modern-day human was not a direct line: the influence of a changing environment on competing, coexisting species led to evolution for some, extinction for others. Today, *Homo sapiens sapiens* is the only surviving hominid line, whose closest living relatives, physically and genetically, are the gorilla and the chimpanzee.

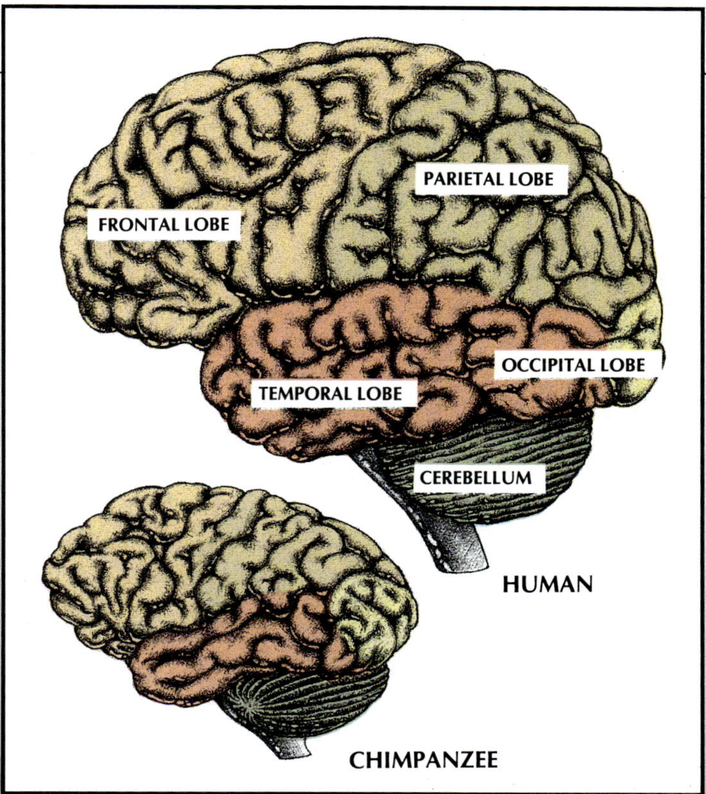

The human brain's superiority over that of the ape lies not only in its size – three times that of the chimpanzee – but also in its shape. The areas responsible for muscular coordination (cerebellum), mental capacities such as intelligence, memory and personality (frontal lobe), interpretation of sensory information (parietal lobe) and hearing and speech (temporal lobe) are all more highly developed in humans. Only the occipital lobe (responsible for vision) is more developed in the chimpanzee.

means of achieving such an adaptation – to circumstances which themselves were constantly changing – was through genetic change.

Surprisingly, this change would be initiated in many cases by an accident. Then, as now, it sometimes happened that the DNA copying process would not be entirely accurate, with the result that the new organism inherited a characteristic or characteristics radically different from that of the original organism. Mostly the changes would be for the worse and the new mutant organism would perish. Sometimes the change made no appreciable difference. But, very occasionally, a mutant creature would be capable of doing some things better than its predecessors, or, in a few instances, of doing things its predecessors just could not do.

For example, a creature would be born possessing a new characteristic – such as a sharper tusk – which, although putting it out of step with the rest of the species, would nevertheless prove to be such an asset that its bearer became more genetically successful than its counterparts: that is to say, it lived longer and bred more offspring than they did. As time wore on, and the success story repeated itself from generation to generation, the sharper-tusked creature became the norm, until it too was replaced by another mutant strain, perhaps one with shorter, more agile legs.

This capacity for genetic improvement was perfectly embodied by our Ethiopian australopithecine, down by the lake's edge. For although her lineage stretches directly forwards to large-brained, manually dexterous modern man, it also stretched back to small-brained, ham-fisted creatures, who stood with all four limbs firmly in the ape camp.

Thirty million years ago, in the tropical rainforests of Africa,

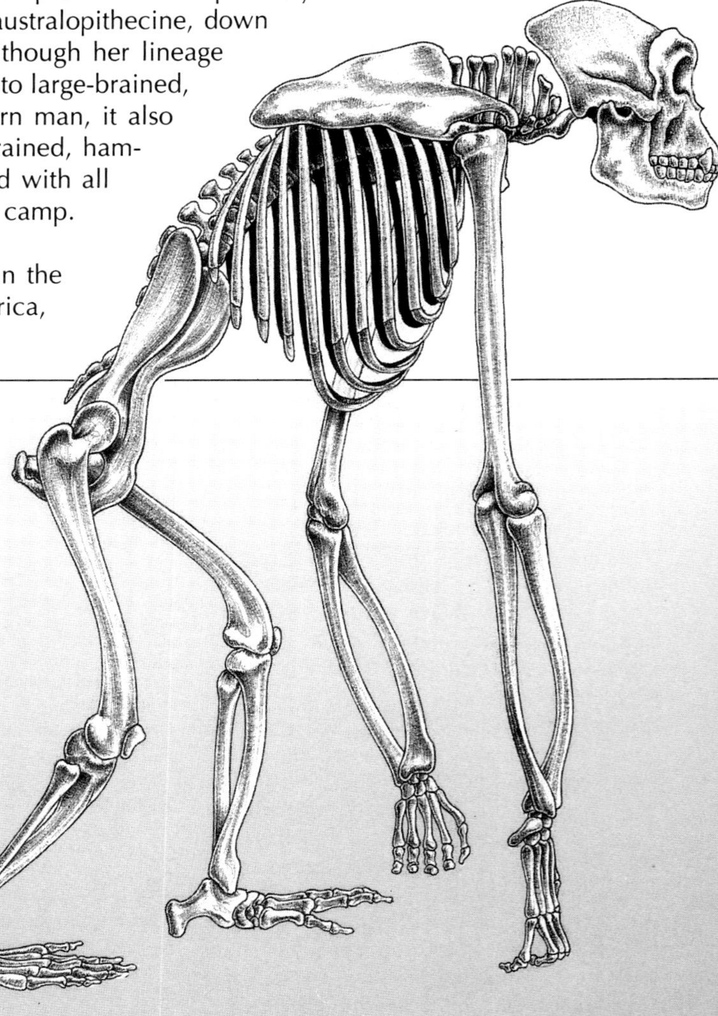

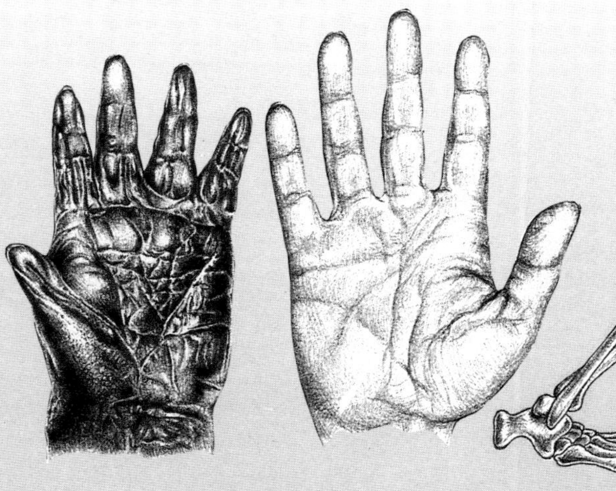

Though the hands of apes and humans display obvious structural similarities, the human thumb is not only longer, but has far greater scope for movement than that of the ape. By contrast, the flat of the ape's hand is larger than that of the human, giving it a more powerful grip but less digital dexterity.

monkeys lived much as monkeys do today. They spent their days high up in the leafy canopy, scampering along the branches and using their long tails to steady themselves when they stood still on all fours, and when they leapt from tree to tree in search of fruit. However, at around the same time, a different breed of monkeys, rather larger in size, took advantage of their longer arm span to initiate a new form of food gathering. Their technique was to dangle. Simply by attaching one hand on to a branch and then lowering themselves in the right direction, they could grab fruit that had previously been out of reach. Gradually, the inventors of this aerial harvesting method progressed from passive dangling to active swinging, in the process developing wrists that were more mobile, and arms that were longer still. And since they now provided their own means of balance, their tails not only became redundant but actually disappeared. Tail-less, they could no longer be classified as outsize, unconventional monkeys. They belonged to a new category altogether. Apes.

Slowly their new life style brought about further changes. Perhaps because of a different diet, requiring more chewing and grinding, apes developed sturdier molars, with five cusps, or points, rather than the monkeys' four. Gradually, too, their posture became more upright than that of the monkeys, and their ability to sit rather than crouch freed their hands for the more intricate tasks of plucking, poking, holding, stroking and examining. This gave them a clear advantage over monkeys, who could grip a nut or a small piece of fruit in their hands, but – if they had to make a quick getaway – needed all four paws for running, and consequently would have to drop their food. Apes, on the other hand, could make off with their prize still clasped in one hand. As an added bonus, it also turned out that the more the apes used their hands, the more intelligent they became; increased manual dexterity stimulated the growth of their brains, which in turn equipped them with the necessary intelligence to perform yet more complex tasks.

And there they – and we – might have stayed, enjoying a comfortable superiority over the monkeys, and swinging contentedly through seemingly limitless forest. But

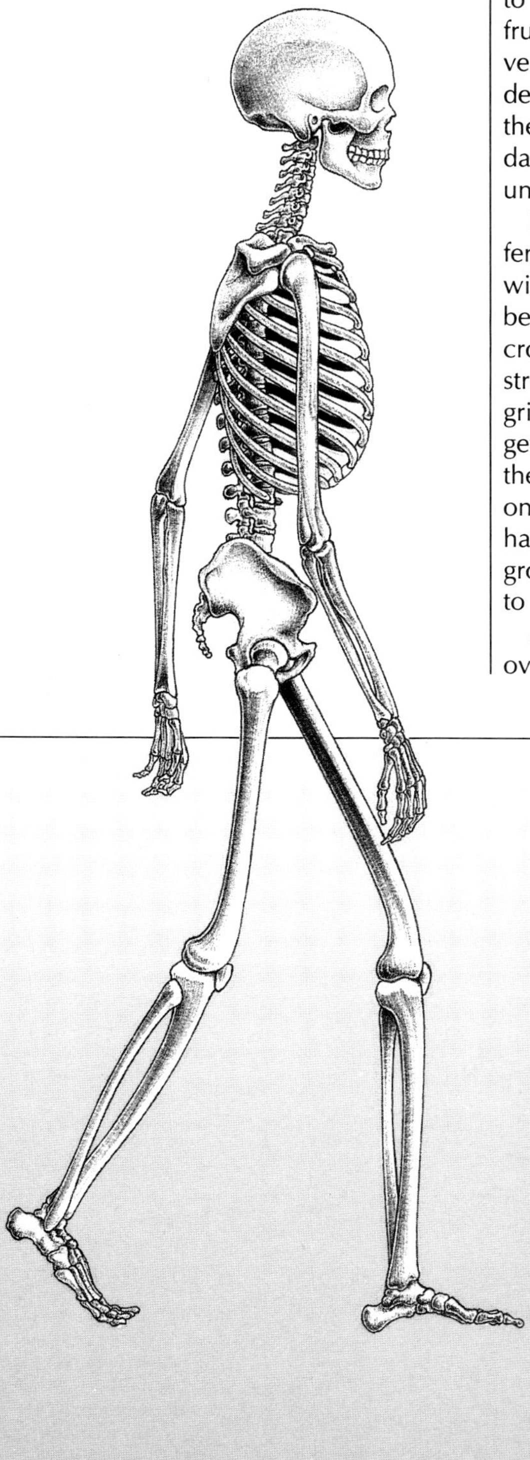

Bone for bone, the human skeleton corresponds with the ape's, but the demands of upright walking produced significant changes in shape. The spine developed a curve, to support the vertical posture; the pelvis became shorter and basin-shaped, to accommodate the stance and gait, and the legs lengthened. Hominid behaviour and diet reduced the need for prominent canines – used by apes in displays of hostility – and caused the gaps between the teeth to disappear.

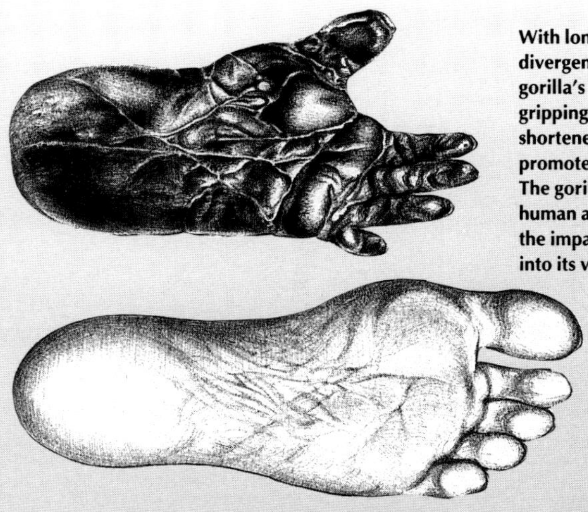

With long toes and a divergent big toe, the gorilla's foot is made for gripping, while the shortened toes of humans promote upright stability. The gorilla lacks the human arch which absorbs the impact of each step into its vaulted structure.

it was not to be. In about 15 million BC, the world's climate began to change, slowly becoming more seasonal and more arid. In the great lush rainforests, where food had always been in season, sparser deciduous trees came to dominate. In many places the forest thinned away to savanna, which was nothing more than open grassland with scattered trees. Suddenly, the animals that hitherto had dwelt within the forest had to adjust to a new set of circumstances. Those who lived on the ground needed only to alter their diet; antelopes and horses, for example, switched from leaves to grass. However, those creatures whose homes had been the trees were forced to make more drastic changes if they were to survive.

The first ape that we know adapted successfully to life away from the forests was *Ramapithecus*, the Rama-ape, named after the Hindu god Rama. The fossilized remains of *Ramapithecus* are found as far apart as Spain and China, but they are at their most plentiful in the foothills of the Himalayas in Pakistan, where 13 million years of erosion from the mountains has left layers of sediment nearly 3,000 metres deep. Among these sediments are preserved the jaws and skulls of several specimens of *Ramapithecus*, between 10 and 12 million years old.

The Rama-ape was of modest dimensions – no taller than 1.2 metres – but what marked it out most clearly from other apes was its teeth. The canines, which in earlier primates had been dagger-like, had in this species shrunk until they were not much larger than the incisors, and the molar crowns had developed thick, hard-wearing layers of enamel. With this new arrangement *Ramapithecus* could move its lower jaw from side to side, unimpeded by protruding canines, and grind its food between molars now strengthened for the task. These were not the jaws of a creature that fed on soft, mushy tropical fruits plucked from the tree tops, but those of an animal that fed on hard objects such as grass seeds and grains. At least one ape, it seems, had come down from the trees for good.

Unfortunately, there is a vast gap in the fossil record that covers the period when human beings' ancestors diverged from the African apes. For five million years after

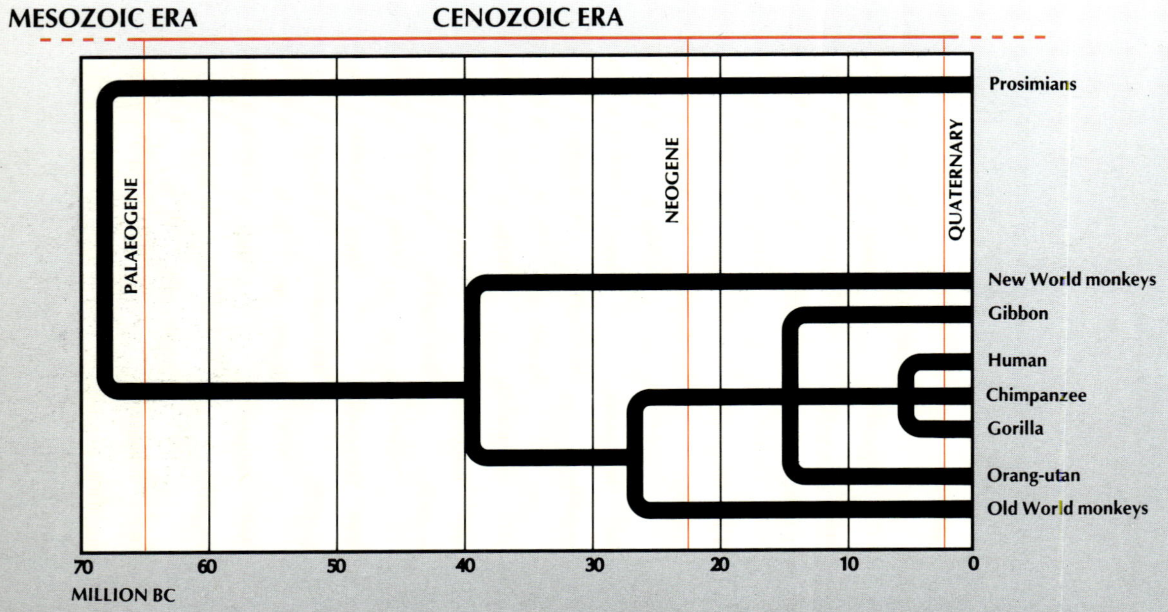

A DNA chart *(left)* comparing chromosome make-up shows the closeness of humans' relationship to the gorilla and chimpanzee, and the comparatively recent date at which their evolutionary paths diverged. DNA – deoxyribonucleic acid – is the main constituent of chromosomes, which are responsible for passing on hereditary characteristics in animals and plants. Human DNA differs by less than two per cent from that of chimps but by nearly eight per cent from that of baboons. The evolutionary development chart *(right)* illustrates the gradual and indirect process which led to modern *Homo sapiens sapiens*. The areas shaded grey indicate dating that is still under debate.

the last *Ramapithecus*, between 10 million and five million BC, almost nothing has survived of the apes. Only the occasional five-cusped tooth has surfaced to show their continued presence in Africa and Asia. When the record is again clearly legible, we find evidence of strange new creatures, no more than 1.25 metres tall, with brains little larger than those of chimpanzees. Their canines were small and their molars were thick, but they possessed one characteristic that was more astonishing than all the rest put together. They walked upright.

These were the australopithecines, or southern apes, whose remains have been found at sites in both South and East Africa. The earliest australopithecines lived in around five million BC and the last disappeared from the earth in around one million BC. The fossils of these first bipeds showed a great deal of anatomical changes from those of their ancestors. The hole in the base of its skull through which the spinal cord passed – known as the foramen magnum – faced almost directly downwards, whereas in apes it faced outwards from the back of the skull. The spine, too, no longer formed a hunched arch, as in other primates at that time; instead, it was a sinuous, S-shaped pillar, erect in the middle, and curving slightly at the neck and in the small of the back. The pelvis had become shorter, broader and more bowl-shaped, so as to support the weight of the torso above. The hip joints were modified so that the legs could lie in line with the backbone, and the knees had grown closer together so as to bear the total body load more efficiently. The ankles had become stronger and more inflexible than an ape's, the soles of the feet had grown vaulted rather than flat, and the big toe had become aligned with the other toes, thereby losing its ability to grasp, but acquiring the capacity to take the full weight of the body for an instant in each stride, as it still does with us today. Examination of footprints made by australopithecines at Laetoli, in Tanzania, shows that their gait exactly matched that of modern-day peoples who have never worn shoes.

Australopithecines varied quite markedly in their physical dimensions. Fully grown adults of the smallest species could measure as little as 1.2 metres in height, and

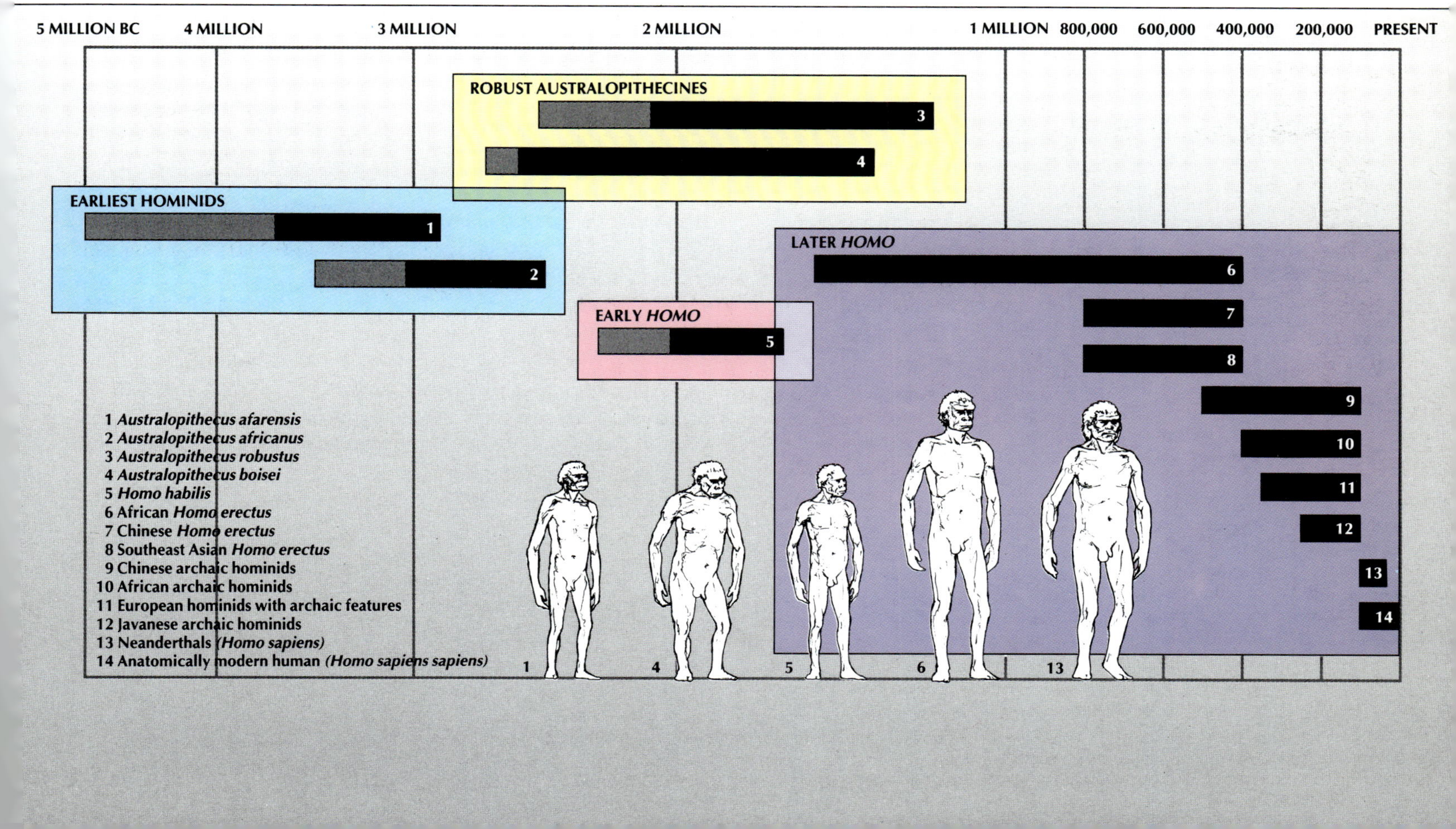

weigh not more than 30 kilograms, whereas the biggest members of the largest species stood up to 1.55 metres tall and weighed 70 kilograms. However, they did possess some notable features in common: all of them had large, well-developed molars, and brains that were much smaller than those of modern-day human beings. The average capacity of an adult australopithecine's cranium – the part of the skull housing the brain – was about 450 cubic centimetres, 65 cubic centimetres more than that of a chimpanzee, but nearly 1,000 less than that of an average person today.

Few clues remain as to the australopithecines' life style. No tools, for example, have ever been found that might indicate the range of routine tasks they performed. Raymond Dart, the Australian-born anatomist who made the first australopithecine discoveries in 1924, originally supposed that these first hominids lived by hunting, because the first bones were found in caves among the remains of antelopes and other middle-sized prey. Later, however, scientists realized that these australopithecines were the prey, too. Far from dying alongside the bones of the beasts they had killed, they themselves had fallen victim to some large animal – perhaps a leopard – which had devoured them and then discarded their bones.

Virtually unchanged from its ancestors of 400 million years ago, a modern-day coelacanth *(immediate right)* drifts through the deepest reaches of the Indian Ocean. Whereas environmental change forced the evolution of new characteristics among most other species, the coelacanth seems to have possessed the inbuilt capacity to adapt to that change without developing radically different features. Other life forms that have undergone similarly few changes over many millions of years include, from left to right, the bat, the turtle, the opossum and the cycad, a palmlike plant on whose cones dinosaurs once fed.

Judging by their large molars, ideal for grinding coarse plant and vegetable foods, the australopithecines seem to have lived as herbivores, perhaps going up into the trees at night for safety, but spending the day down on the grasslands and lakesides of Africa. They shared their environment both with the ancestors of animals that still live there today, and with some – such as the giant grass-eating baboon, more than 1.8 metres tall – that have since become extinct.

Most scientists agree that there were four main species of australopithecines. A representative of the earliest species – *Australopithecus afarensis* – has already been mentioned, in the re-created scene by the Ethiopian lakeside, where she and her companions lived around 3.5 million years ago. The American anthropologist Don Johanson discovered her fossilized remains in 1974, when a flash flood washed away the walls of a gully to reveal a piece of her arm bone. The next species for which we have firm evidence is *Australopithecus africanus*, the first remains of which were discovered in even more dramatic circumstances – with the aid of gelignite.

One day in 1924, at the Taung limestone quarry near Johannesburg, the day's blasting was continuing as usual when, after one explosion, workmen inspecting the site came across two most unusual pieces of rock. Embedded deep in one of these pieces was half a skull which, judging by its size, was that of a small child, with its milk teeth still in place. Set into the other piece was the fossilized cast of this same child's brain, with all its convolutions and furrows still clearly visible. Sensing immediately the significance of their discovery, the workmen took the rocks to the quarry manager, who then sent them on to Raymond Dart. After using – among other implements – his wife's knitting needles to pick away the lumps of surrounding sediment, the anatomist carefully examined the small skull and declared that it belonged to a member of a hitherto unknown species. He named it *Australopithecus africanus*, the southern ape of Africa.

Later discoveries of similar fossilized remains – also in South Africa – gave a clearer picture of what *Australopithecus africanus* looked like. Somewhat taller and

stronger in build than *afarensis*, *africanus* inhabited the earth between three and two million years BC. It stood upright and, although not very tall – 1.3 metres at the most – nevertheless possessed a sturdy and muscular physique and weighed somewhere between 45 and 50 kilograms. Compared with what is known about the life style of *afarensis*, *africanus* seems to have enjoyed a more specialized plant diet. Its molars had become much larger and its jaw had increased considerably in strength, in order to facilitate a more efficient grinding action. However, its canines and incisors – the teeth that are at their most developed in carnivores – had grown smaller. This development indicates that the types of food eaten by *Australopithecus africanus* required little preliminary tearing or slicing, and probably just consisted of a variety of fruits or seeds covered by hard pods or husks that needed to be ground down before giving up their contents.

The trend towards plant eating continued in the two later australopithecine species, *Australopithecus robustus* and *Australopithecus boisei*, both of which appear to have lived between two million and one million BC, before becoming extinct. Remains of the former were unearthed in South Africa in 1938, and it certainly deserved its name: it stood no more than 1.55 metres tall, but a big male may have weighed as much as 70 kilograms. Both its skull and jaws were very strongly built, with massive bone ridges that served to anchor the powerful chewing muscles that its diet required. These developments went even further in *Australopithecus boisei*, whose remains were discovered in 1959 at Olduvai Gorge, in modern-day Tanzania, by the archaeologists Mary and Louis Leakey. This species was named after Charles Boise,

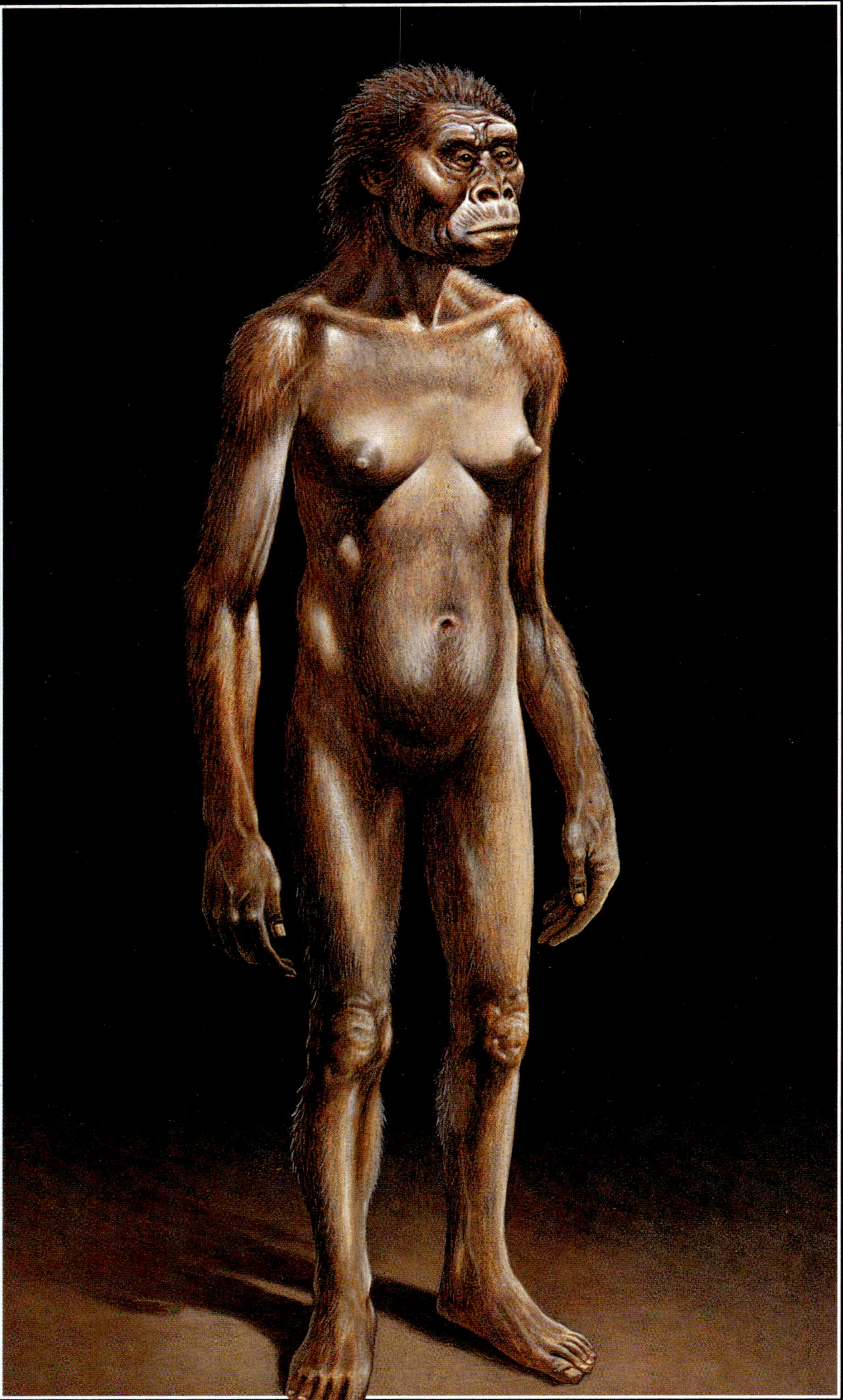

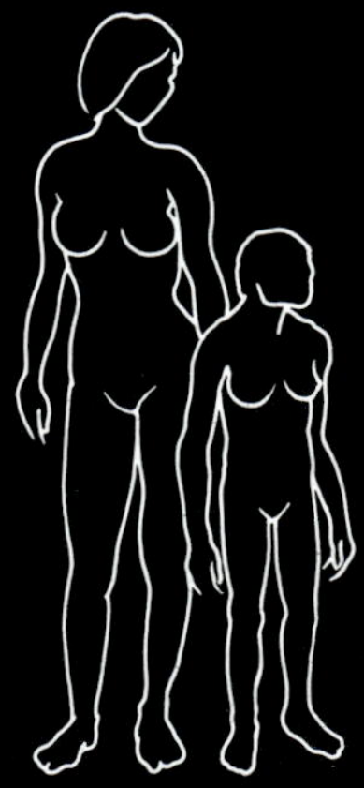

In this artist's reconstruction, "Lucy", a female hominid of the species *Australopithecus afarensis*, gazes out from the distant past of around 3.5 million years ago. Her skeleton was found in 1974 by the anthropologist Don Johanson at Hadar in Ethiopia, and she was named after a song – *Lucy in the Sky with Diamonds* by the Beatles – that was played over and over as Johanson and his colleagues celebrated the discovery. Lucy, standing about 1.2 metres tall and weighing around 32 kilograms, was smaller than a woman of today *(above)*. Although she was not human, Lucy had many human-like features, most notably her erect stance and her hands.

the American-born London businessman who had funded the work undertaken by Mary and her husband. So vast had this specimen's jaws become that their owner was popularly dubbed "Nutcracker Man".

No-one questions the fact that what unites the australopithecines with modern-day human beings is their ability to walk upright. But what no-one has yet agreed on is why they should have chosen to do it.

There are obvious disadvantages. In biological and biomechanical terms, walking on the hind legs with the body held vertical is an extraordinarily complicated and difficult balancing act. Also – if judged on speed alone – it is a blatantly inefficient way of moving, since it makes use of only two limbs, whereas quadrupeds such as apes and monkeys – with four limbs at their disposal – can easily outdistance a human being. In most mammals each limb makes a contribution to the locomotive force, but in human beings the arms are a dead weight. The only bipeds that can move quickly are those such as ostriches and kangaroos, whose front limbs have become so small and weak that they are no impediment to running.

However, one major advantage that bipeds do have over quadrupeds is that they possess far higher levels of endurance. Whereas human beings in large numbers are capable of running a 42-kilometre marathon, a zebra will fall down exhausted after being chased for only about 800 metres.

So what could have caused australopithecines to favour two-footed locomotion as a means of getting around on the ground? Other apes and monkeys have come down from the trees and yet have remained quadrupeds, standing upright only briefly in order to get a better view or to reach a high fruit. Why not the southern apes?

Several factors must have combined to make upright walking the most efficient solution to the problems that they faced. One element may have been the australopithecines' need to present less of their bodies to the fierce tropical sun; an upright creature receives only two thirds the amount of sunlight as one on all fours, and even less when the sun is directly overhead.

Another contributory factor may have been an increased use of the hands, which could have come about for two reasons. The first of these is that the earliest primates communicated by a primitive system of calls, and therefore needed to devise a different method of carrying food and other objects than in their mouths. The second possibility is that they needed to keep their hands free for performing manual tasks and for transporting food back to their fellows. Whereas apes today live in rather loose family groups, in which each individual ape feeds itself, the australopithecines might have evolved a system whereby they brought back the food they foraged to share it with other members of their family. If so, then possessing a good carrying ability would have been vital for survival, and the most successful australopithecines would have been those able to master the combination of efficient portering (using their two upper limbs) with swift running (using their two lower limbs).

That they achieved this in significant numbers is demonstrated by our existence today. Most experts tend to agree that modern-day human beings are descended from australopithecine stock, although they largely disagree over the nature of that descent, as indeed they disagree over the relationship between the various species of australopithecines themselves.

How, when and where the changeover occurred from australopithecines to human beings – members of the genus *Homo* – is still far from clear. What we do know, however, is that *Australopithecus* and *Homo* existed alongside each other in Africa for a substantial amount of time, perhaps for as long as 800,000 years. Far from the doomed and brutish creatures that they are often portrayed as being, the australopithecines at first seem to have achieved rather greater reproductive success than did their hominid counterparts: in excavations carried out at various African sites, archaeologists have unearthed almost twice as many fossils of australopithecines as they have those of the genus *Homo*.

Contrary, then, to popular legend, the apemen did not suddenly die out with the arrival of human beings, nor did human beings deliberately wipe out their more primitive forebears. Those australopithecines that did not follow the evolutionary path towards humankind became extinct solely because they failed to adapt to the increasingly competitive situation on the African savanna, where they found themselves overtaken firstly by more efficient herbivores – such as the gazelle, the oryx and the horse – and then by a growing number of species, including early *Homo*, that had incorporated meat into their diet. Squeezed on the one hand between these more efficient herbivores, and on the other by more efficient omnivores, the last of the australopithecines lost his battle for survival and finally disappeared from the face of the earth in around one million BC.

But, while the australopithecines were on the evolutionary way down, there emerged on the earth, in around 2.3 million BC, a new genus of creature that was very much on the way up. He was man: not an ape or even an apeman, but the world's first fully fledged *Homo*, a meat-eater and grass-eater combined, who not only had a brain that was more than 50 per cent bigger than that of any australopithecine, but who also possessed an ability that would permit him to make his mark on the environment, and shape his daily existence, in a way that no creature before had been able to. He could make stone tools.

This was *Homo habilis*, "man the skilful", whose remains were first discovered, like those of *Australopithecus boisei*, at Olduvai, in eastern Africa. Judged by his height and weight he was not particularly remarkable; no more than 1.25 metres tall and 50 kilograms in weight. But he had an 800 cubic centimetre brain and he was capable of doing what no australopithecine is ever thought to have done: he split stones in order to make sharp-edged cutting tools. He probably did the same with wood as well, but unlike his stone implements these have long since rotted, leaving no traces, and we can only guess that a creature with the intelligence to exploit rocks for his own ends must have done the same with trees.

From the cut marks on the bones of large mammals that have been found at *Homo habilis* sites, we can tell that one of his primary uses for these tools was to butcher animal carcasses. However, this does not mean that he could get hold of unlimited amounts of meat. Analysis of these bones shows that the parts to which *Homo habilis* had access were usually the least fleshy parts of the carcass. Often, too, he seems only to have got the leftovers; many of the tool marks run over the top of the much bigger incisions made by large-toothed carnivores, who presumably had already helped themselves to the choicest cuts of meat.

It seems certain, then, that *Homo habilis* had added meat to his diet, perhaps motivated initially by the dwindling of vegetation stocks, because of either seasonal shortages or competition from other animals. It also seems clear that *Homo habilis* obtained his meat not by confronting large beasts in hand-to-hand combat, but by a policy of cautious and opportunistic scavenging. This involved either picking up the scraps that larger creatures had left behind, or else using sharp tools to open up the corpses of animals whose hides were too thick for other carnivores to pierce. This latter method at least gave the hominids first bite at the carcass before being chased off by large predators, such as lions and panthers, or by any lesser animals, such as hyenas, that could muster in sufficient numbers.

Meat in the diet brought social as well as nutritional benefits. An animal carcass represented a substantial amount of food, usually more than one individual could manage, and it is likely that a pattern of behaviour emerged in which *Homo habilis* shared feasts with his fellows – if only because he could then expect favours in turn. Although this hardly amounted to an organized system of food distribution, this partaking of communal meals must have provided these hominids with an opportunity for social activity and for the forming of a rudimentary sense of fellow feeling towards members of their own kind.

The pursuit of meat also expanded geographical frontiers. Whereas most plants could only grow in a place that offered the right conditions – light or dark, hot or cold, moist or dry – animals were, on the whole, less sensitive to the niceties of shade and temperature, and consequently were inclined to roam farther afield, drawing their human predators after them.

This increased level of exertion, beneath the tropical sun, may have been a contributory factor in bringing about human beings' relative lack of body hair. It seems quite possible that hominids chose to do their hunting and scavenging during the hottest times of day, when other larger predators were resting; repeated over many thousands of years, this practice could have led to a gradual shedding of body hair and an increasing dependence on sweating as a means of keeping body temperature down in the midday sun. For human beings perspire more profusely than any other mammal, even though our sweat glands are proportionally the same size as theirs. And whereas other carnivores use bursts of panting as a method of cooling themselves, this is only effective after short bursts of exertion; over the kind of timespan that scavenging hominids might need to chase a large animal till it collapsed, panting would have led to dangerous hyperventilation.

Sweating provided hominids with the most suitable evolutionary solution to high-temperature living. However, along with the benefits that it brought, it also imposed a number of restrictions. The steady loss of body liquid meant that hunters had to confine their operations to areas where they knew that drinking water would be readily available. The human frame will not tolerate a water loss of more than 10 per cent of its total body weight, and whereas camels, for example, can stock up with large amounts of liquid, swallowing 100 litres in 10 minutes, human beings can manage only one litre in the same time. Closeness to water therefore became essential to those engaged in active pursuit of meat.

Finding ways of meeting these new challenges both stretched and stimulated the hominids' intelligence. Whereas for as long as three million years, the brain of an average australopithecine had remained the same size proportionally as that of a chimpanzee, the human brain now embarked upon a slow and continuous process of enlargement that would end only when it was more than three times bigger than that of the largest australopithecine.

Several complex and interrelating factors brought about this growth. Scavenging and hunting made more demands on the intelligence than eating grass, especially in a species that did not possess the physical assets of a killing machine such as, say,

The Taung skull, uncovered in South Africa in 1924 and named after the limestone quarry in which it was found, was the first firm evidence of the presence of early hominids in Africa. It belonged to an australopithecine child of about six years old. Raymond Dart, professor of anatomy at Witwatersrand University, quickly recognized that the skull must be hominid from its shape – too high and rounded for an ape – and from its full set of milk teeth, which are scarcely larger than those of a human child. Subsequent skeletal finds of *Australopithecus africanus* suggest that had the Taung child been male and grown to adulthood, it would have stood about 1.3 metres tall and weighed 45 to 50 kilograms.

the leopard or the lion. And access to a food like meat, that was rich and high in protein, provided human beings with the additional energy source they needed to fuel the development of a bigger brain.

The making of tools required a level of dexterity that also served to promote an increase in brain size. No-one precisely understands the correlation, but it has been observed that animals exhibiting a degree of manual skill, such as the chimpanzee, are usually more intelligent than those that do not display such an ability. We can assume, therefore, that the very act of chipping stone to make tools must have literally broadened *Homo habilis'* mind, providing him with the intellectual capacity to perform yet more complex manual tasks, which in turn would have had a further nourishing effect on his brain.

Tool-making also demanded a mental input that went well beyond that of merely knowing how to make the implement. Visualizing the existence of a sharpened edge within a rounded stone was an act that required a new level of imagination and insight. Also, while *Australopithecus* merely needed to know where to go to find food, *Homo habilis* had to know where to go to find the raw materials for his tools as well. This kind of activity necessitated foresight, memory and planning, in a way that plucking berries off a bush did not.

As the size of the human brain increased, so the head grew in order to accommodate it. Here, though, a problem presented itself. For the new upright walking position limited the width of the human pelvis; if it got any wider, the legs would become splayed and inefficient, thereby setting human beings back on the road to quadrupedalism. However, this in turn limited the width of the female birth canal, thereby restricting to about 350 cubic centimetres the maximum cranial capacity of a newborn baby, who had to pass through that canal during birth. Since most primates' brains only doubled in size between infancy and adulthood, this therefore meant that the brain size of a fully grown adult would never get past 700 or 800 cubic centimetres. At this point, then, it seems as if the evolutionary conveyor belt had ground to a halt. What had previously been human beings' inexorable progress towards a bigger-brained future was blocked by what was a seemingly insuperable obstacle. Either bipedalism would have to go – or brain growth.

However, the forces of evolution were still at work. Gradually, and unconsciously, human beings began to evolve a changed pattern of development. Mothers started giving birth to children whose brains were to do most of their growing outside the womb – a process that has reached a point today whereby the average human brain at the time of birth is only 25 per cent of the size it will reach as an adult; this compared with 65 per cent for a chimpanzee. The advantage of this new arrangement lay in the fact that hominids' brains now quadrupled in size during their lifetime; there was one disadvantage in this change, though, in that throughout the period that the growth of the brain was at its most intense – during the first few years of life – their body growth became retarded.

The price that humankind paid for this increased brain capacity in later life, therefore, was an increased helplessness at the start of it. Whereas a newborn horse could stand up on its own and walk within two hours of birth, and a baboon could look after itself at 12 months, human babies became utterly dependent on their mothers for anything up to the first six years of life.

This called for a completely new social structure. Previously, female primates had mated, and then reared their children unassisted. Now the sexes began to cooperate in nursing and bringing up their helpless young, an arrangement that in turn required the setting-up of more stable and complex relationships between individuals.

Some of our distinguishing human characteristics can reasonably be supposed to have evolved as a result of – and as a response to – these social interactions. The reproductive cycle of the human female, which keeps women sexually receptive for most of the time, instead of just during the mating seasons, may well have developed both out of the community's need for parents to form stable bonds, and out of the greater personal intimacy between male and female that was engendered by that stability; so may have the human female's ability to experience orgasm, unparalleled in any other species. It has also been suggested that sexual fidelity between men and women came about because their upright posture allowed them to mate face to face, the resulting eye contact having a bonding effect that was denied to those animals that were obliged to mate front to back. However, this theory is not supported by the example of the orang-utan, an ape that mates face to face but is neither bipedal nor even approximately faithful.

Judging from the locations in which his fossilized remains have been discovered, *Homo habilis* appears not to have strayed outside the boundaries of East Africa; finds in both France and Pakistan of what are claimed to be tools made by him still remain unsubstantiated. But even though *Homo habilis* may not have felt prompted to find out more about the outside world, evolution did not stand still, and in around 1.6 million BC *Homo habilis* gave way to a more advanced race of hominids, who were known as *Homo erectus*, or upright man.

These new hominids were directly descended from *Homo habilis*, but were built on a larger scale. Their biggest representative would have had as much as 50 centimetres height advantage, and weighed 20 kilograms more than even the bulkiest *Homo habilis*. Possibly these new hominids had grown so much larger as an unconscious response to food shortages. For the bigger an animal gets, so the lower its energy requirements become in proportion to the size of its body; a marmoset, for instance, needs proportionally three times as much energy as a human being, and therefore has either to find supplies of high-energy food – such as nuts or meat – or else spend virtually the whole day eating low-energy foods, such as grass and leaves. Faced, therefore, with an environment in which high energy food was scarce but low-energy materials abounded, hominids might well have evolved larger frames in order to accommodate this imbalance.

It was not only their bodies that had grown. The average *Homo erectus* brain now measured 950 cubic centimetres, as against *Homo habilis'* 800. With this improved mental and physical equipment, the new hominids took the art of tool-making far beyond the mere chipping of rocks. From the right piece of stone, *Homo erectus* now had the ability to make symmetrical, pear-shaped implements known as hand axes, with which they could more easily butcher whole carcasses of meat. They had also learnt how to make stone scrapers – for jobs such as cleaning flesh off hide – by knocking small flakes off a larger block.

At some point they also discovered how to use fire. The first traces of this are equivocal: at Chesowanja in East Africa, researchers discovered pieces of burnt clay, which were dated to 1.4 million BC. Animal bones and stone tools found nearby show that early hominids – probably *Homo erectus* – had been present, and tests on the clay indicated that it had been heated to 400 degrees Celsius, a temperature typical of a campfire blaze. But it is difficult to be sure; other specialists point out that the clay could have been burnt in a bushfire.

There is no definite evidence for the controlled use of fire by humans throughout the next million years. Caves at Zhoukoudian, near Beijing, have yielded human remains from about 460,000 years ago, along with layers of what was originally thought to be burnt charcoal, but what has now been identified as burnt owl and bat droppings. Again, it is not possible to tell for certain whether these droppings had been burnt deliberately by human beings, or whether, in fact, they had merely been consumed by naturally occurring fire.

The exact date at which *Homo erectus* acquired the gift of fire remains unknown, therefore. What we are sure of, however, is that *Homo erectus* – unlike *Homo habilis* – spent his days not just in the balmy climes of Africa but also in the chillier zones of Europe and China. If, as seems probable, he crossed over into these colder continents from his native Africa, fire would have been essential to him in securing a foothold. Without it he certainly would never have been able to survive a winter on the exposed and windblown steppe of eastern Europe.

Not that the opening part of the Quaternary period, from two million to 8000 BC, was particularly hospitable anywhere in the world. Periodic variations in the earth's orbit, along with the constant shifts taking place in the shape of the world's continents, combined during this time to produce a climate that was perpetually fluctuating, alternating approximately every hundred thousand years between ice ages and warmer intermediary periods known as interglacials.

During the interglacials, such as the one that we are enjoying today, the equatorial regions experienced an abundant rainfall, while the northern and southern parts of the world experienced relatively temperate climates. However, each interglacial would gradually grow colder, and the ice would spread southwards until it covered large parts of Europe and North America. Northern Asia became bitterly cold, but lack of rain would keep it free of ice. Then, after a few thousand or sometimes just a few hundred years, this glacial period would reach its peak and quite suddenly be replaced by a new interglacial.

Throughout this alternation of climates, *Homo erectus* doggedly survived, changing remarkably little either physically or mentally over a period of some 1.5 million years. He learnt to make slightly improved tools, and he mastered the ability to make twig huts, but his biggest strides forward came in hunting. By around 300,000 BC, bands of nomadic *Homo erectus* hunters were tackling – and killing – animals as large as elephants, and were using fire to drive big game into enclosed areas where they could more easily be slaughtered.

They moved around in bands of about 20 to 30 individuals; Africa, Asia and Europe together might have supported 40,000 such groups – fewer than a million people across the whole globe. How they lived is a matter for conjecture. Probably the complex vocal articulations of human beings today were beyond them, and in any case, coordinating a band of hunters to bring down large prey can be done by gesture and force of example; wolves manage it quite efficiently.

And so *Homo erectus* continued, making his own slow, unspectacular way along the evolutionary path. However, soon after 300,000 BC, that path seems to have deviated. We know this from remains found all over Africa, Asia and Europe, which around this time start exhibiting a mixture of physical characteristics: they have the larger brains and thinner skulls of a modern *Homo*, but they still retain the heavy

Shown in the drawing above as it appeared nearly two million years ago, Olduvai was the site of a lake surrounded by lush vegetation and a wealth of animal life. As streams from nearby volcanic mountains slowly silted up the lake, hominid habitats were buried. Geological movement later shifted the drainage centre east and a stream began to carve out a 15-kilometre gorge. The erosion of this now arid landscape *(left)* uncovered vital clues to human evolution.

A CRADLE FOR EARLY LIFE

Olduvai Gorge in Tanzania has yielded, through the dedicated work of a husband-and-wife team of archaeologists, a unique insight into the practical and cultural world of hominids between two and 0.7 million years ago. Louis Leakey, a pioneer of the search for human origins in Africa, was attracted to Olduvai by its abundance of fossils and animal bones, which led him to believe that it might also have been home to hominids. After spending almost 30 years uncovering thousands of fossils and stone artefacts, he and his wife Mary were, in 1959, rewarded by the discovery of a 1.75-million-year-old hominid skull. They nicknamed this first fossil evidence of *Australopithecus boisei* their "Dear Boy". Then, two years later, they discovered *Homo habilis*, the earliest known species of *Homo*, who was subsequently credited with producing the older stone artefacts that they had found, and hence called "handy man". The later tools in the Gorge are the work of *Homo erectus*. After Louis Leakey's death in 1972, his work in Africa was continued by his wife, and by his son Richard.

Tools found at Olduvai include a primitive pebble chopper *(right)*, 1.8 million years old, and a more sophisticated 800,000-year-old hand axe *(left)*.

Homo habilis had a longer skull than early hominids – as in this drawing of a male head – allowing for a larger brain.

bones and the flat, rather than high-domed, foreheads of an altogether more primitive man. These remains were of hominids who came to be classified as "archaic *Homo sapiens*". Some of these seem to have gone on to evolve naturally into anatomically modern man, or *Homo sapiens sapiens*. Others – particularly those found in Europe and the Levant – underwent a quite different kind of transition. They became, by 100,000 BC, Neanderthal man, a creature whose name has since become – most unfairly, as it so happens – a byword for the epitome of primitive brutishness. Scientists used previously to classify this beetle-browed hominid as an entirely separate species – *Homo neanderthalensis* – but he has now been firmly instated as our very close relative *Homo sapiens neanderthalensis*, whose cultural and technological achievements are discussed in the next chapter.

Compared to their predecessors, the Neanderthals were to hold only a limited tenure on the earth, for by 30,000 BC they had all but disappeared. Whereas *Homo erectus* had existed for more than a million years and had made remarkably few innovations, the Neanderthals vanished after just 70,000 years, during which time they had, among their many other achievements, revolutionized the art of tool-making to the point where it could be classed as mass production.

The creatures that replaced them were startlingly different in appearance. Although no more than 20 centimetres taller than the average Neanderthal, they were infinitely more slender and nimble, and weighed between 30 and 45 kilograms less. There were changes in the features, too: above their eyes rose tall, elegant foreheads – uninterrupted by heavy Neanderthal brows – that continued first upwards and then sloped back downwards, forming the domed cranium that housed the brain. This was *Homo sapiens sapiens*, "man the double wise", a human being who was anatomically indistinguishable from ourselves. He first appeared in South Africa around 100,000 BC, and within the next 70,000 years had replaced all the previous species of hominid right across the entire globe.

Precisely how he achieved this remains one of the enduring mysteries of the human story, and one that plays a central part in the next chapter of this book. That *Homo sapiens sapiens* came to dominate the rest of the world with his technology and culture is unquestioned, but what no-one knows is how he achieved this in what is, in biological terms, a minuscule amount of time.

Possibly it was because he had at his disposal a tool that was to prove more useful than any of his bone sewing needles, a weapon more powerful than any of his stone spearheads. That is to say, language.

With language *Homo sapiens sapiens* could pass on technological knowledge. With language he could mastermind complex cooperative ventures, such as what appears to be an organized seaborne migration to Australia in around 50,000 BC. And with language he could give expression to abstract thoughts and concepts that were manifested in a great outpouring of paintings, drawings and carvings that began in Europe in around 30,000 BC.

He unquestionably had all the vocal equipment for it. His pharynx had grown proportionally much longer than that of earlier hominids, and his tongue had an increased flexibility that enabled him to form and emit a much wider range of

Squatting in silent contemplation, this *Homo erectus* male grips a pear-shaped hand axe in his strong fingers. Generally viewed as the rootstock of modern *Homo sapiens sapiens*, *Homo erectus* had limbs and torso similar to modern man. The pronounced brow-ridges, deep-set eyes and protruding jaw, however, proclaim his descent from *Homo habilis*, the earliest tool-maker. The axe – probably for chopping, cutting and pounding meat and plants – is one of the simple, versatile tools that appeared soon after the emergence of *Homo erectus*, around 1.6 million years ago.

sounds in quick succession. But he paid a price for this increased vocal range; his longer pharynx meant that the larynx had descended so low in the throat as to join up with the alimentary canal, thereby creating a risk of food accidentally getting into the windpipe and causing him to choke.

Homo sapiens sapiens' improved speaking ability may have had some effect on his outer appearance, too. This notion stems from observation of human babies today, who are all born without a fully developed pharynx. At the age of three months, as their vocal tract starts to take shape, their larynx drops down in their throat; at this point, the base of their skull becomes arched, forming a roof over the pharynx. It is possible that over thousands of years this arching upwards of the skull-base had the effect of pulling the outwardly thrust *Homo erectus* face inwards. The inward pressure would then have forced the brain backwards and upwards, leading to the rounder-domed *Homo sapiens sapiens* skull we possess today.

Although not proven, this theory would account for the sheer speed of the biological changes that transformed *Homo erectus* and his descendants into anatomically modern men. The innumerable advantages that a speaker would have possessed over a nonspeaker add up to a powerful evolutionary reason for such rapid change. Nonspeakers, or those with a restricted range of language, would have become virtual second-class citizens, relegated to the fringes of the habitable world, condemned to living only where the speaking population did not want to live and to eating only what the speaking population did not want to eat. Very soon, their inferior living conditions would have meant that their numbers began to dwindle, while those of the speaking population increased.

Much of this, of course, remains speculation, based on archaeological finds in the Near East, Africa and Europe. Hardly any evidence exists for what was happening at the time in Arabia, central Asia or India, and no-one can rule out the possible existence of fossils just waiting to be exposed by a chance flood or quarry blast, which may shed new light on the mystery: a Neanderthal still alive in 10,000 BC or a *Homo sapiens sapiens* in Europe before 100,000 BC.

For the moment, though, the existing evidence points to *Homo sapiens sapiens* having carried out, perhaps with the aid of language, what amounts to an evolutionary coup d'état. By 30,000 BC, anatomically modern human beings had achieved what amounted to a supremacy over all other species on the planet. From this point onwards, the story of mankind's development ceased to be that of nature shaping human beings and began to be that of human beings shaping nature.

LANDSCAPES OF THE ICE AGE

When the last ice age reached its zenith around 16,000 BC, average temperatures in parts of Europe and North America stood as much as 18 degrees Celsius lower than they are now. Between a quarter and a third of the entire land surface of the globe lay entombed beneath a shroud of ice in parts three kilometres deep.

Much of this mass was in motion. In places, spectacular cliffs of frozen water some 100 metres high formed the spear-head of advancing ice sheets, progressing between 600 and 1,200 metres a year, gouging and tearing up the ground as they went, grinding huge boulders into powder or else picking them up like marbles and dumping them hundreds of kilometres away from their source.

In the northern hemisphere, the ice travelled south from the Arctic Circle, thrusting out in a number of different directions. North America had two giant ice sheets: one came down the east coast, blotting out eastern Canada and New England, and then heading inland across the Midwest until it eventually reached the point where St. Louis, Missouri, stands today, 450 kilometres south of Chicago. The second ice sheet extended down the west coast from the Rockies, engulfing parts of Alaska, all of western Canada and parts of Washington, Idaho and Montana. Meanwhile, in Europe, an ice sheet stretched out from Scandinavia, swallowing up most of Britain and Denmark and large parts of northern Germany, Poland and the Soviet Union.

Because the southern hemisphere consisted largely of open ocean, less land was lost to the rapacious invader. In tropical Africa some glaciers did manage to form on the highest ground, but it was only in the Andes in South America, and in the mountains of New Zealand that this process reached spectacular proportions. Meanwhile, the vast Antarctic ice sheet thickened and spread north.

Because so much of the earth's water was locked up in ice, radical changes occurred in the planet's geography and climate: sea level dropped by 125 metres, exposing vast areas of what had once been sea bed; the English Channel and the Persian Gulf disappeared; in Southeast Asia a huge new continent emerged, linking the islands of Indonesia in one large landmass, and Siberia and Alaska were united by a broad land bridge. Rainfall levels dropped, too; average rainfall in Britain was the same as in modern-day Morocco. Winds blew fiercer and stronger than before, scattering huge quantities of dust, rock debris and vegetation.

The pressure of the ice, allied to the increased hostility of the climate, forced the earth's vegetation belts much closer to the equator than they are today, as demonstrated by the map on the facing page. At latitudes where dense forests now flourish, arctic tundra stretched out bleak and featureless; similarly, regions that today are deciduous woodland, were in 16,000 BC treeless, grassy steppe.

Four of the most common environments prevalent at this time are shown on the following pages. In each, human beings and other animals were compelled to adapt their life styles to the resources available. In the process they developed strategies that enabled them to weather the full force of the ice age in all its varied and daunting manifestations.

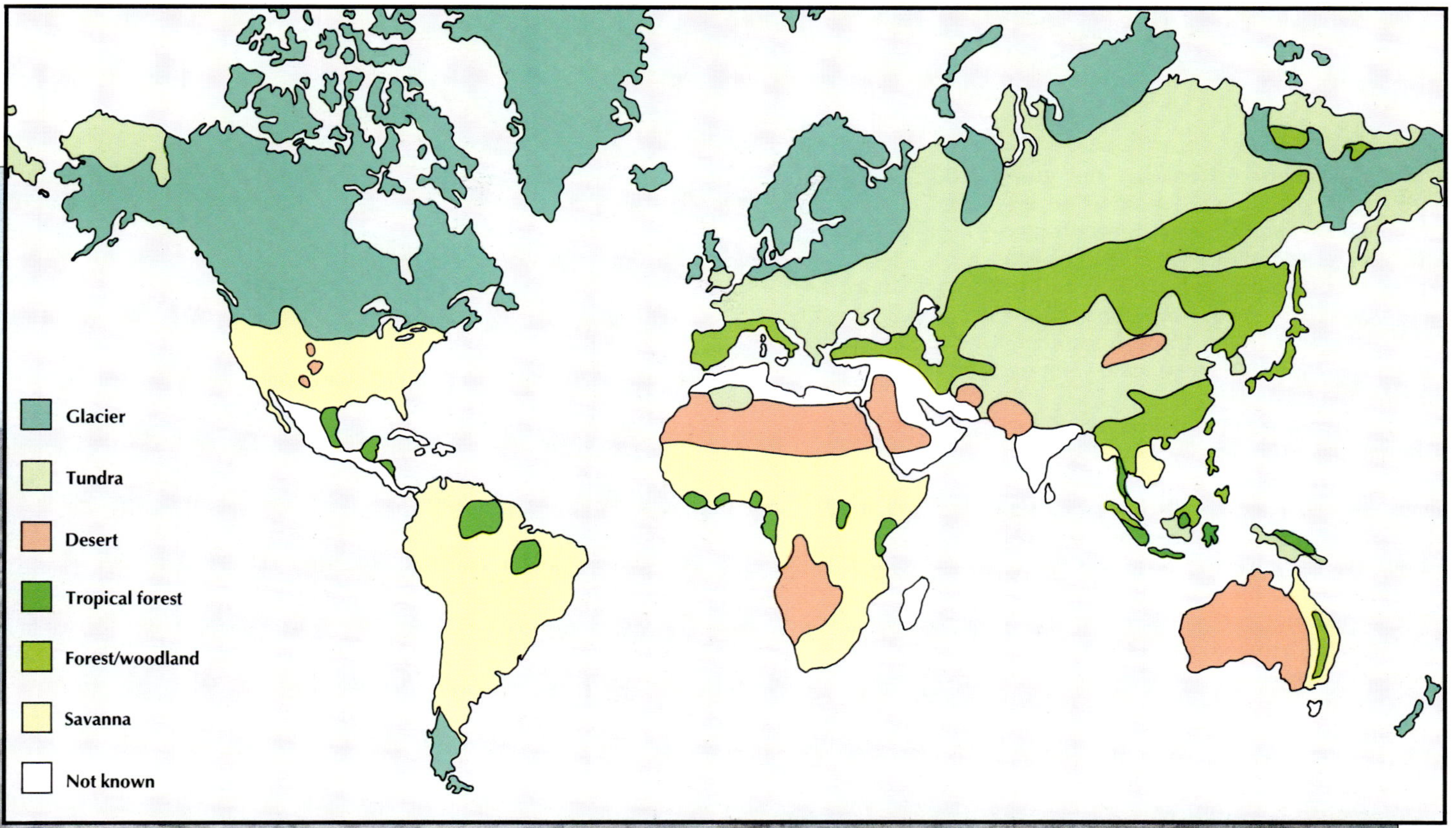
Glacier
Tundra
Desert
Tropical forest
Forest/woodland
Savanna
Not known

THE CHEERLESS TUNDRA

At their southern extremities, the great northern ice sheets gave way to an environment of tundra – a flat, semifrozen zone. In North America the area of tundra was smaller than it is today, but in Europe and Asia it stretched from the Atlantic coast of France right through southern Germany, eastern Europe and Siberia, finishing in China.

Fiercely cold winters melted into short, cool summers, when mosses, heathers and arctic flowers briefly carpeted the great plains. However, no tree could take root, for the tundra soil remained permanently frozen just beneath the surface.

Further south, the tundra merged with areas of warmer and grassier steppe, to which large herbivores such as bison, mammoths and the giant deer – which had an antler span of more than three metres – would migrate in winter, along with horses, deer and the two-horned woolly rhinoceros.

Most human beings did not venture very far into the tundra outside the summer months. However, some of the hardier, or hungrier hunters did brave the howling winter winds to go in search of those animals that could survive the cold, among them the shaggy musk ox and the elusive arctic wolf.

Musk ox

Woolly rhinoceros, now extinct

Collared lemming

Giant deer, now extinct

FRINGES OF THE FOREST

During this period no woodland grew north of the Mediterranean. To the south, however, oak, beech, walnut and chestnut trees spread a graceful canopy across a broad sweep of the world that stretched from the southeastern part of North America, through southern Europe to south China and Japan.

Few human beings chose to live in the forest itself, preferring instead to settle at the edge of a wooded area, ideally near a river or sea. The trees provided a rich supply of fruits and berries in season, as well as fuel and materials for shelter all year round. Caves offered a refuge both for humans and for other hungry predators, which in the woodland environment of southern China included tigers, black bears and hyenas. Other large but strictly herbivorous occupants of the woods were the giant panda and the stegodon, an elephant-like creature whose tusks grew so close together that its trunk could not fit between them.

Those hunters who did go into the forest brought back wild pigs, and small deer such as the muntjac. They also gathered chestnuts, walnuts and any other fruits and seeds that had escaped the attention of the energetic macaque monkeys.

Stegodon, now extinct

Giant panda

Spotted hyenas

Macaque monkeys

THE ARID ZONES

Paradoxically, hot deserts flourished during the ice age, because the cold weather inhibited evaporation – and consequently rainfall. Temperatures near the equator remained high, and the largest arid zones were to be found in the Sahara and Kalahari areas of Africa, as well as in Arabia, northwest India and Australia.

In the regions that were most parched, desert dwellers could live only in those places where they had easy and continual access to water, whether in the form of permanent natural springs, or of water holes that filled up during periods of heavy rain, or of seasonal streams that were fed by melting glaciers in nearby mountain ranges. They survived, like the desert wolves and jackals, by hunting the herds of gazelles and wild asses that migrated through their territory.

In the less barren areas, often near springs or marshes, enough grass grew to sustain the large, grazing aurochs, or wild ox. Men and women too could usually find enough grass seeds, roots and tender plant stems to keep themselves alive. Occasionally they might even have managed to supplement their basic diet with some protein, in the form of ostrich eggs or even the flesh of a small animal such as a tortoise or a Cape hare.

Aurochs, now extinct

Gazelles

Ostrich

Tortoise

Cape hare

Sugar glider

Koala bears

Extinct *Thylacoleo* with young in its pouch

Lakes were highly prized by the people of the last ice age. They guaranteed a supply not only of fresh water, fish and crustaceans, but also of the animals and water birds that gravitated naturally to the lake and could be killed for their meat while they bathed or drank.

However, in around 16,000 BC a global drought was causing these desirable lakeside habitats to dry up. The process had reached an advanced stage in Australia, where it was presenting problems both for human beings and for the animals with which they shared the continent. Many of these creatures formed a part of the human diet: hopping mice, wallabies, kangaroos, koala bears, the echidna, or spiny anteater, and the sugar glider, a flying opossum that carried its young in a pouch. Human beings also hunted a number of larger beasts, including a three-metre-high giant kangaroo called *Procoptodon* and a distant, rhinoceros-size relative of the burrowing wombat, known as *Diprotodon*. In turn, humans themselves would certainly have been wary of the stalking *Thylacoleo*, a fierce marsupial lion about the size of a modern-day leopard and with massive, fearsome jaws.

Giant kangaroo, or *Procoptodon*

Echidna

THE PEOPLING OF THE EARTH

2

On a chilly summer's afternoon, beneath a pale, watery sun, a small band of some 30 people – men, women and children – might have been seen advancing slowly across the bare and cheerless tundra of eastern Siberia. The men, armed with fearsome, stone-tipped spears and protected against the biting wind by tight-fitting costumes of animal hide, walked at the head of the party. As they moved they kept a nervous lookout for bears and other predators, noting any likely hunting spots as they went, and scanning the ground for droppings or hoofprints indicating that reindeer or mammoths had passed that way.

Towards the back of the group came the women, picking their way carefully round the low bushes and lichen-encrusted rocks that littered the ground, and squelching through the clumps of sodden moss and grass. They stooped as they walked, bowed down under the weight of the small children that they were carrying, and of the rolled-up lengths of animal fur and hide which they would later use to build their shelters for the night. The older children ran and played around the party as it progressed slowly across the plain towards the area of higher ground that could be seen ahead, stretching away to the range of low hills lying on the distant horizon. Like the men, the women and children kept a watchful eye on the terrain they crossed, constantly on the alert for anything it might have to offer in the way of food – seeds, nuts and berries, as well as animals.

Who these people were we cannot say with certainty. What we do know, though, is that between 50,000 and 13,000 BC they – or a group very much like them – made this same journey at a period in the world's development when Asian Siberia was joined to American Alaska by a land bridge some 80 kilometres long. And at one point in their journey across this bridge, the travellers would unwittingly have stepped over a nondescript patch of scrub or stones that could be said to constitute the dividing line between the two continents. In so doing, they became the first human beings ever to set foot in America.

Not, of course, that this imagined band had any concept of such momentous implications. As far as they were concerned, they were merely progressing a few more weary and unremarkable kilometres across the seemingly endless and windy steppe, in search of a new and more fertile homeland. It had been the gradual depletion of their food resources, as well as the increasing competition from bands of eastward-travelling migrants, that had forced them to leave their ancestral hunting grounds in Siberia and to make their way east, too, taking advantage of the brief weeks of summer, and the milder weather, to make their move. They had always been a nomadic people, shifting regularly from camp to camp across their old territory, and making the best use of the many different resources it offered at different times of the year. They were accustomed to travelling light, and the tools that they carried

The face of *Homo sapiens sapiens*, clearly modern in form, stares from beneath sharply engraved lines representing braided hair. Excavated at Brassempouy in France, this tiny head – measuring less than four centimetres high – was carved from the ivory tusk of a great woolly mammoth at some time between 30,000 and 20,000 BC. By this time, early humans had spread out of Africa across the temperate zones of the globe, and were developing artistic and technological skills to match their territorial expansion.

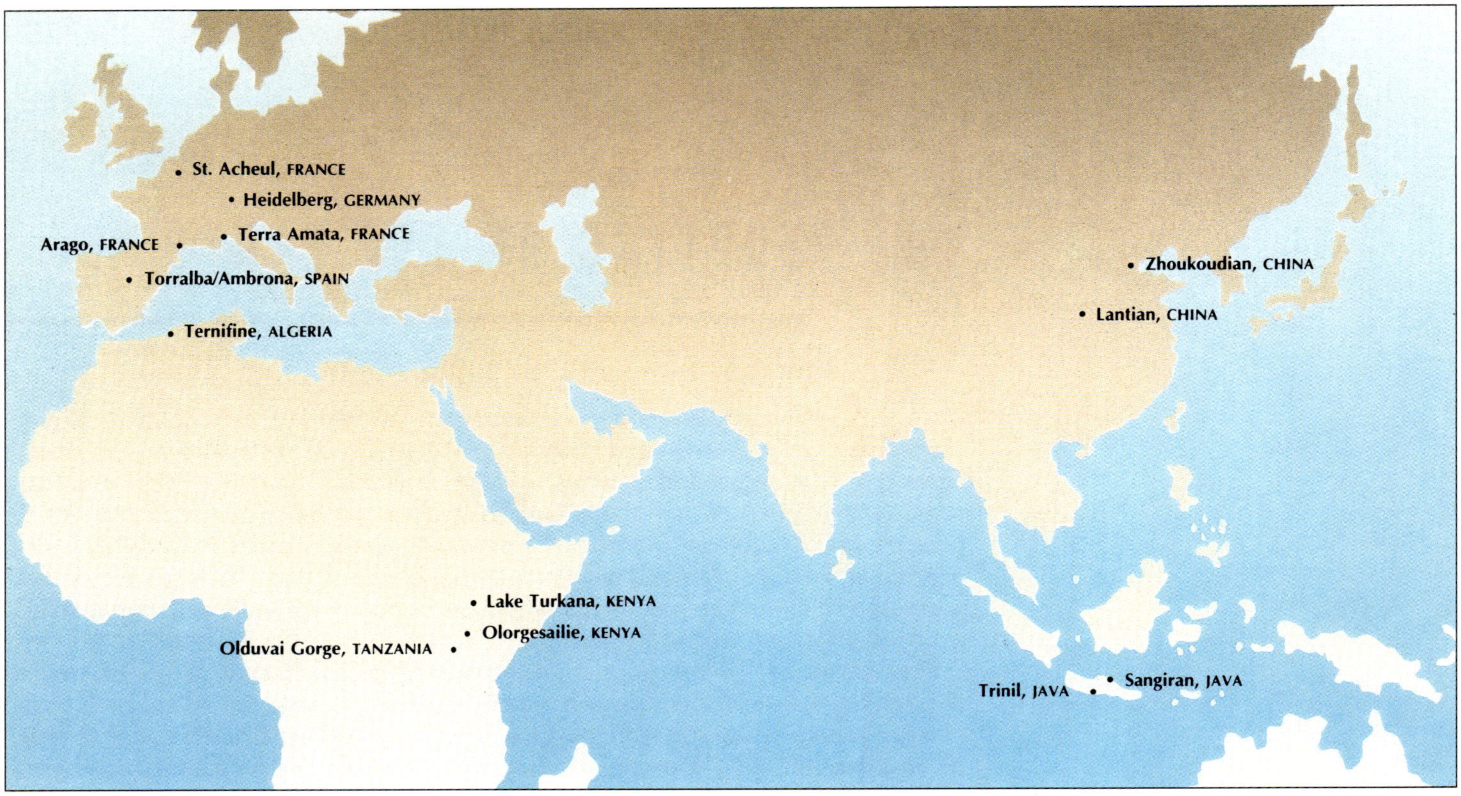

This map shows the main sites at which fossilized bones of *Homo erectus* and other remains such as artefacts and butchered carcasses have been found. The third identifiable hominid, *Homo erectus* marked an intermediate stage between the more apelike *Homo habilis* and the modern human *Homo sapiens sapiens.* Emerging in East Africa some 1.6 million years ago, the species had by 800,000 BC reached Java. In Europe the earliest skeletal traces – such as the Arago skull and fossilized Heidelberg jaw – date from around 400,000 BC, but remains of human artefacts indicate a presence not long after one million BC. The mastery of fire brought safety from predators and promoted new forms of cooperative social organization, speeding the expansion of *Homo erectus* across Eurasia.

with them were simple and disposable, such as rough stone flakes that they could quickly shape and sharpen by skilled chipping. Their prized possessions were similarly few: a handful of figurines, along with a variety of small ornaments carved from bone, ivory or stone. Even their more permanent implements – such as delicate bone needles, sharp flint blades set into handles made of bone or wood, and finely honed borers that were used for piercing hide and engraving bone or horn – could all be quickly bundled into bark containers or hide shoulder bags when the time came for striking camp and travelling.

Now, the gently rising land did indeed seem to be growing richer. In the distance, reindeer were grazing. Fanning out and sweeping round so that they approached the herd downwind, two of the men, together with an adolescent boy learning his hunting skills, silently stalked a young buck that was feeding apart from its companions. The animal, measuring over a metre tall at the shoulder and weighing more than the three hunters combined, continued to graze unsuspiciously. Stealthily, one of the men fitted a spear into the wooden throwing stick that rested on his shoulder. He paused for a short moment, and then, with a sharp flipping action dispatched the spear. His aim was true; the buck pitched forwards, wounded. Still wary, the other men moved in upon the crippled animal, mindful of the flailing hooves and of the lethal pair of antlers. Again and again they darted in to stab the dying beast at close quarters, until at last it lay still.

Quickly now, as dusk was drawing near, they set about the task of butchering the animal. For this they used a mixture of implements – flint chips, which came smooth and sharp, as blades for skinning, and much heavier stone choppers for smashing the thicker bones and more resistant sinews. Their work done, the three rejoined the main body of the party, their shoulder pouches full to bursting with fresh meat. That night, beside the newly constructed hide shelters, reindeer steaks sizzled over the camp fire, and well-fed children slept deep.

These first Americans undertook their long journey towards the end of a two-million-year-long period during the course of which both human beings and the world that they inhabited experienced a series of great climatic shifts – long, protracted freezes followed by sudden thaws. Enacted over tens of thousands of years, these changes were sometimes slow, sometimes perceptible within the span of one human life, but over a number of generations they radically altered whole environments, locking the seas and rivers in great ice sheets that subsequently thawed, increasing sea levels and submerging large tracts of land.

As well as geographical schisms, the early Quaternary witnessed the opening of a definitive genetic gap between humans and the apes. This would leave the bipedal australopithecines firmly on the ape side, where they would die out. It also established *Homo habilis* – "handy man" – as the first indisputable member of the genus *Homo*, later to be joined by his larger and more sophisticated counterparts, *Homo erectus*, "man the upright", *Homo sapiens*, "man the wise" and finally anatomically modern man, or *Homo sapiens sapiens*, "man the double wise", who in evolutionary terms is indistinguishable from us today.

The physical difference between *Homo habilis* – 45 kilograms of ape-jawed primitive – and *Homo sapiens sapiens*, who dressed in a suit and tie would pass for a 70-kilogram sales executive, was immense. But although he was more advanced genetically, in both cultural and technological terms *Homo sapiens sapiens* inherited a great deal from his less refined ancestors. It was they who had acquired fire, who had devised ever more elaborate methods of tool-making, and who had over thousands of generations evolved increasingly complex patterns of work and social life, thereby giving the basically weak and defenceless human species a distinct competitive edge over all other animals. These skills ultimately enabled humans not only to gain mastery over rival species and to cope with environmental and climatic changes, but also to colonize the entire globe, from the blazing hot tropics right up to the frozen wastes of northern Siberia.

From the very beginning, right back as far as 2.3 million years BC, *Homo habilis* had developed a taste for meat. It was a food that had considerable appeal in terms of nutrition as well as flavour, because compared with fruit and vegetables, meat is a highly efficient way of taking in nourishment. One kilogram of venison, for instance, provides almost six times as many calories as the same weight in berries or green vegetables. Thus the securing of one deer provided its captors with the same amount of nourishment as would many hours of backbreaking foraging on the hillsides and scrublands. Also, meat-eating gradually destroys the organisms used by the stomach to break down the cellulose in vegetables, which consequently become less easy to digest; over a number of generations, this must have been a contributory factor in humans following a more carnivorous diet.

Nevertheless, killing large wild animals was no easy task, and strong though the first hominids' cravings for meat may have been, they must often have gone unsatisfied. Australopithecines, for example, had been neither swift nor sharp enough, nor had they developed sufficiently advanced tactics to hunt game successfully.

Homo habilis was considerably better placed physically, with a far more developed manual dexterity, and a brain that was almost 50 per cent larger. With these assets he developed ways of stalking and killing smaller animals (as troops of modern-day baboons, working cooperatively, have been observed to do), and of driving larger ones into swamps or over cliffs where, once cornered, they could quickly be dispatched with a hail of rocks. Such rudimentary hunting was, however, a highly opportunistic and unsystematic affair, which meant that for *Homo habilis*, as for his australopithecine predecessors, scavenging still remained the main means of acquiring meat with any degree of regularity. Therefore, apart from those occasions when they had the good fortune either to surprise their prey or else to come across a beast that was weak or injured, *Homo habilis* hunters were obliged to scavenge what meat they could from the carcasses of animals that had either been half-eaten and discarded by larger carnivores, or had died of natural causes. Often, especially when very hungry, the early hominids must have had to drive away from the carcass other feasting animals, such as lions and leopards – a distinctly hazardous business if one

AN EARLY EDEN

Shown here as it would have appeared around 400,000 years ago, sheltered in a deep river valley, this site of a hunter-gatherer encampment at Bilzingsleben, in modern East Germany, commands a richly diverse environment. As the diagram on the opposite page demonstrates, each type of vegetation was home to a variety of animal and plant species, which together satisfied all the material needs of the human community that settled here. The rhinoceros, for example, the most common hunted animal, provided not just a plentiful supply of meat and bone marrow but also the large bones used for building shelters and for tool-making. The fur of beavers, deer and other animals was used to make clothing.

is just 1.25 metres tall and armed with nothing more substantial for protection than a stick, or at best a stone tool. This being the case, the diet of *Homo habilis* must have remained largely a vegetarian one, although now and again they may have been able to supplement it – as chimpanzees do – with termites that they had winkled out from a nest using a long straw or twig.

However, as time wore on, the later hominids changed physically, and as they changed, developed an increasing capacity to adapt to their environment and to alter the world around them. They might have lost the ability of their predecessors, the tree-dwelling primates, to grip with their feet, but this was more than made up for by the fact that their hands were becoming ever more sophisticated instruments. Whereas all primates could seize objects with their hands and let them go at will, only the hominids developed the distinctive "opposable thumb", which, by pressing back against the other gripping fingers, could allow the delicate handling of objects and an infinite number of precise digital adjustments. In this respect the hands of *Homo erectus* scarcely differed from our own, and this manual dexterity allowed the

Wide swathes of open pasture were roamed by herds of bison, deer and horses.

Deciduous woodland sheltered elephants, aurochs, wild pigs and deer, and also yielded an abundance of nuts, berries, herbs and other edible plants.

Scrub was home to rhinoceroses and elephants; roots, berries and nuts were also plentiful.

Marshes were inhabited by water birds and waders, nesting among the reeds and stalks.

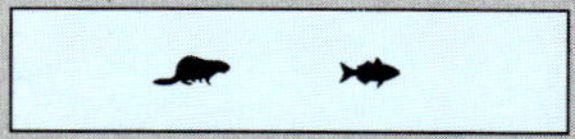

Rivers, streams and pools were stocked with a variety of freshwater fish; colonies of beavers lived on the margins of land and water.

Bears, and predators such as lions, wolves and foxes, roamed freely on land from one type of vegetation to another.

later hominids to produce stone tools for chopping and cutting that far surpassed the crude creations of other primates.

In tool-making, the most distinctive product of *Homo erectus* was the hand axe. These pear-shaped tools of rounded triangular profile, with a point at the top and a broad curved butt, remained one of the most important tool types in Europe, Africa and western Asia from 1.5 million BC up to about 100,000 BC. Why they continued to be used for so many millenniums is still something of a mystery, but they clearly demonstrate that *Homo erectus* and his immediate successors were conservative in their tool-manufacturing, and were content to go on creating similar shapes of stone tool for hundreds of thousands of years.

Another major contribution to technological progress was *Homo erectus*' mastery of fire. Scientific opinion varies about how this was achieved. Some say that hominids first created fire by accident, from sparks produced by rubbing together sticks or stones. Others say they literally captured the flames by dipping sticks into the edge of either forest fires or volcanic lava flows.

Whatever the truth of the matter, there can be no doubt that fire immeasurably improved the quality of hominid life. It provided warmth, of course – something that was becoming increasingly important outside the tropics as the ice ages bit deep. Indeed, it is very likely that the colonization of Europe and Asia by *Homo erectus* that began around a million years ago was only made possible by the knowledge of fire, both as a tool to ward off predators and as a means of keeping warm in the fierce cold of the northern winters. Other species had to depend on the slow process of genetic change – such as the development of a thicker, warmer coat – to adapt to colder climes, but man could rely on his growing intelligence and ability to make clothes, and to build fires and shelters to keep himself warm.

Fire could also be used for cooking meat. The advent of cooking constituted a significant breakthrough in more ways than one. It rendered meat not only more appetizing, but also more digestible and beneficial, because heat breaks down some of meat's more complex compounds and releases a variety of nutritious juices. Eating cooked meat meant that *Homo erectus*' teeth grew smaller, since they were no longer required to bite through large chunks of raw flesh and gristle; and the introduction of softer, cooked food also removed the necessity for the large, obtrusive jaws that humans had needed to assist them in what must often have been the epic task of chewing the raw flesh once it had been bitten of. Some indication of what meal times must have been like 1.5 million years ago can be gleaned from a *Homo erectus* jawbone found near Heidelberg in Germany, which has cut marks clearly visible on the front teeth. Evidently *Homo erectus* ate meat by clamping it between his teeth and then hacking at it with a sharp flint tool.

However, the most fundamental contribution that fire made to improving human beings' lot was its capacity to inspire fear in other animals. For while the hominids had over millions of years become meat-eating predators themselves, they had not developed anything like the natural defences that were needed to fight off rival carnivores such as lions or sabre-toothed cats. Such predators could wreak havoc among the hominid groups, and parties venturing from their shelter at night were liable to be torn to pieces by predators that had picked up their scent. Such vulnerability receives graphic testimony from remains found in South African caves. The skull of one unfortunate australopithecine, dated 1.5 million BC, was found in a cave at Swartkrans, halfway between Johannesburg and Pretoria, scarred by the deep and unmistakable toothmarks of a leopard. The big cat apparently took its prey into a tree to eat, out of reach of competitors, and at the end of the meal dropped the discarded skull into a nearby cave shaft.

Fire, however, kept all such predators at a distance. The importance of the flame is underlined by the fact that some groups of hominids kept their fires burning – or at least smouldering – all the time, both inside and outside their caves or shelters.

The fire thus became the centre of the community's activities, the focus for a far more complex social life than had yet been seen. The idea of a home base, around which a community could organize its life more cooperatively and efficiently than when permanently on the move, was taking root for the first time.

Along with this idea of establishing a home base went the development of shelters. As *Homo erectus* spread from his tropical homeland to colder regions in Europe and Asia, he relied on his cultural skills to adapt to the harsher climate. The discovery of fire gave him a major asset, but without clothing or shelter the colonization of the temperate latitudes would have been impossible. In some areas, caves provided convenient natural shelters; but where these were lacking, *Homo erectus* was forced to improvise with the materials to hand and build shelters of his own. One of the earliest is the camp site of Terra Amata, on the Mediterranean coast of France, near Nice. Here archaeologists discovered the remains of 11 small oval huts, built of branches that rested on a central row of upright posts. Around the edge of the hut the bottom of the branches were held in place by heavy stones. Situated near the centre of each hut was a hearth which, together with other items that have been found, give us a vivid and quite detailed picture of what life was like in this early European settlement. The occupants sat and slept on animal skins, and may have painted their bodies with ochre. A faint trace in the sand showed where a wooden bowl had been placed; a scatter of flint chips reflected the efforts of a tool-maker. From analysis of the bone remains we know that the people of Terra Amata ate red deer, elephant, rhinoceros, goat and boar; the shells of oysters, mussels and limpets also unearthed, show that the seashore was not far away.

As part of his social development, *Homo erectus* probably took further than his predecessors the sexual division of labour, which had developed alongside the evolving human brain. As hominid brains had gradually increased in size, so it had become biologically necessary for baby hominids to be born at earlier – and more helpless – stages in their growth.

The demands of constant child care restricted the movement of women, and the presence of clinging, chattering young certainly hindered them in the stalking of easily alarmed prey. So the task of ranging far and wide, scavenging and hunting for meat, devolved on to the males, while that of foraging closer to home for vegetable food and gathering it in became the province of the females. The "hunter-gatherer" life style, as it is known, was a formula that was to survive for almost a million years, during which time *Homo erectus* became firmly established outside Africa, across Europe and Asia, and in China, Java and other parts of the Far East. It continued for such a long time that it was even inherited by the different subspecies of *Homo sapiens* to which *Homo erectus* gave rise.

Of these, the most famous today are the Neanderthals, so called because the first remains of one of their number were found in West Germany's Neander Valley, near Düsseldorf, in 1856, when workmen at a limestone quarry dug up a skullcap, ribs and fragments of pelvis and other limb bones. The owner of the quarry thought that

the bones were those of a bear. The local schoolteacher was of the opinion that they belonged to someone who had been washed down into the quarry during the biblical Great Flood. A German anatomist confidently suggested that the bones were those of a Russian Cossack who had fought against Napoleon in 1814 and, after deserting, had crawled into the cave and died. For almost a hundred years the word "Neanderthal" carried connotations of clumsiness and stupidity, an opinion that was based solely on the appearance of these bones. The myth was reinforced by the discovery 50 years later of the first complete Neanderthal skeleton, at La Chapelle-aux-Saints in southwestern France, which seemed to indicate that Neanderthals had had a stooping posture and a slouching walk. Now, however, it has been discovered that the shambling, brutish appearance of the man from La Chapelle-aux-Saints was due not to any natural awkwardness, but to his having suffered from arthritis.

Contrasting forms of shelter demonstrate the ways in which early humans adapted to different climatic environments. At Zhoukoudian in China, generations of *Homo erectus* between 460,000 and 230,000 BC lived permanently in deep limestone caves such as that shown above, warmed by fires that also provided heat for cooking and security from wild beasts. In the gentler climate of southern France, their contemporaries at Terra Amata made do during summer hunting trips with light, temporary shelters built from leafy branches strewn upon a rough framework of saplings *(opposite page)*.

Previously, many people had shied away from the idea of Neanderthal man being one of our direct ancestors, for he presented an outer appearance that did not harmonize with the popular image of "civilized" man. Above his eyes travelled an overhanging brow-ridge of solid bone, topped by a long, flat skullcap that broadened out into a bulging, bunlike cranium at the back and sides. The lower half of his face thrust out dramatically from his neck and contained large front teeth which he would put to a variety of practical uses: chewing animal hide in order to soften it, or clamping pieces of wood so that his hands would be free to carve or to chop them. Stubby feet, short legs and a massive, barrel-chested frame weighing anything up to a hundred kilograms completed the picture of a muscle-bound throwback.

Physically, they made both their hominid predecessors and their modern human successors look like weaklings. Their broad, flattened noses provided large nasal cavities in which inhaled air was warmed before entering the lungs, their thrust-forward face structure keeping this cold air well away from the brain. Their bones were more solid – some 10 to 20 per cent heavier – than those of their modern successors, and the great muscles required to control them produced further heat. For the whole of their history, between 100,000 and 30,000 BC, the Neanderthals experienced all the freezing rigours of a full-blown glacial climate. That is one reason that they developed highly sturdy, robust frames.

The maintaining of body warmth was not, of course, the only reason for the Neanderthals' mighty physique. Wherever they found themselves, even in such temperate regions as the Middle East, the Neanderthals led a tough existence. Their nomadic way of life was a punishing regimen involving, it has been estimated, walking more than 30 kilometres a day over rough and frequently icebound terrain, often carrying heavy loads. Having strength in the upper body was essential, and the Neanderthals' handgrip was probably two or three times as powerful as our own.

And yet, for all their rugged, brawny appearance, the Neanderthals were the match of later humans when it came to brain capacity. Whether this meant they could match them in intelligence is not certain, but what does emerge clearly from excavations of Neanderthal sites is that they took the art of tool-making to new levels of refinement. Previously, a tool made by any of the earlier hominids – even the talented *Homo erectus* – had essentially been a shaped stone, a rock painstakingly whittled away to produce an often cumbersome implement. But the Neanderthal workman devised a way of producing two or three neat and virtually identical flakes from a piece of rock that would earlier have afforded only one. He used a technique called the prepared core method: first, the craftsman knocked the top off a round rock, much like slicing the top off a boiled egg, which created a flat, disc-shaped surface or core. He then worked his way round the edge of this disc, striking downwards at the core with another rock to produce a series of flakes, which he could then fashion into knives or hand axes by setting them in wooden handles.

As well as introducing this small – though nonetheless crucial – element

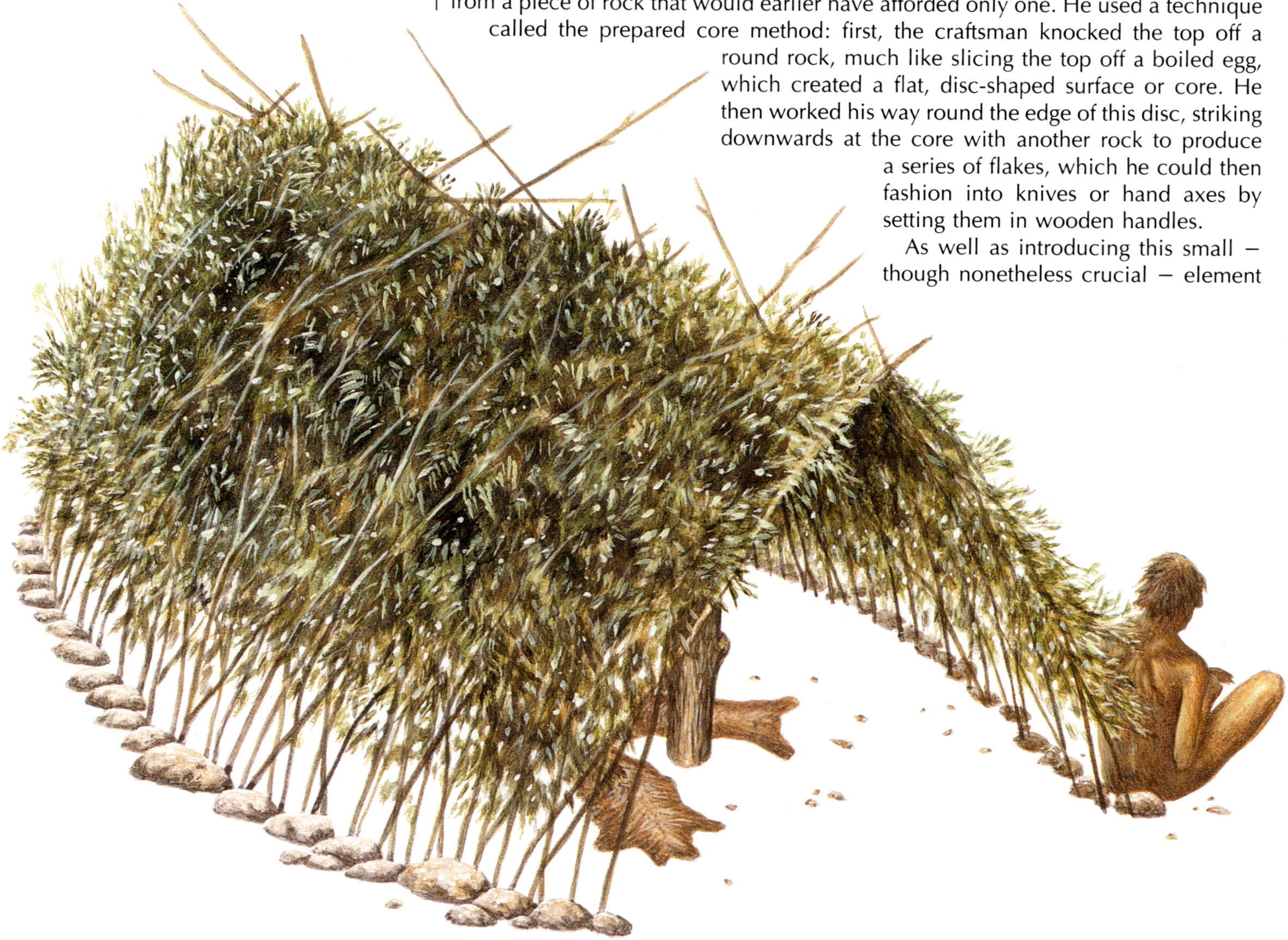

of mass production to the manufacture of tools, the Neanderthals brought a greater organization and efficiency to the business of hunting. Large-scale, preplanned game drives became a regular feature of Neanderthal times; horses, mammoths, and other large mammals were herded into muddy, swampy areas, where they could be more safely attacked at close quarters. Large quantities of meat – beyond the wildest dreams of the earlier hominids – could be secured in this way. Whatever was surplus to immediate requirements would be stored in special pits dug in the ground; and because permafrost began less than a metre below the surface in many parts of Europe, the meat placed in these natural refrigerators would stay good almost indefinitely. This bulk consumption of meat did not necessarily make for any lack of refinement, however; small scratches found on Neanderthal teeth would seem to indicate that these giants probably used toothpicks after a meal.

The cultural life of the Neanderthals also outshone that enjoyed by their ancestors. Some groups seem to have honoured particular species among the animals that they knew, and some made a practice of preserving in special shrines the skulls and bones of creatures that they killed. The cult of the bear, for example, seems to have been particularly widespread: one cave in Switzerland was found to contain 13 carefully arranged bear skulls, some of which had been packed inside a rough chest made of heaped-up stones, others set in little niches in the wall. Further evidence of this cult was unearthed by archaeologists excavating a Neanderthal site in France: underneath a stone slab that weighed almost a tonne, was discovered a rectangular pit containing the bones of more than 20 bears.

Little is known about what actually happened at such shrines. A cave in Lebanon contains evidence of a fallow deer having been ritually dismembered and sprinkled with red ochre, presumably symbolizing blood, but for the most part we can only speculate on the precise nature and function of Neanderthal ceremonies. What we

In an illustration based upon evidence discovered at Wyoming in the United States, spear-toting hunters of around 8000 BC close in on a herd of bison that they have backed up against a sand dune. This hunting technique was a refinement of the early hominid ploy of chasing prey into terrain that formed a natural trap. Once penned in, herds of animals started to panic, making it easier for the hunters, operating in groups of two or three, to capture a beast which in normal circumstances they would not have dared confront.

can say with a degree of certainty, however, is that the lives of the Neanderthal people contained a religious content.

We suspect, for example, that Neanderthal people believed in some form of afterlife. The food, tools and animal skulls that have been found buried with Neanderthal corpses all seem to suggest that there was a widespread belief in the existence of a world beyond the grave, where these items would prove practically useful or provide some kind of magical protection in the next world. At La Ferrassie, in France's Dordogne, what appears to be the site of a family cemetery has been excavated. Here a number of bodies were found lying side by side: a man, a woman, a child of around five years old, and two babies. All the corpses had been carefully arranged in shallow, parallel trenches oriented east-west. Flakes of flint had been buried with the man, and a flat tablet of stone placed over his head and shoulders, intended either to protect him or to prevent his ghost from leaving the grave. Some way off a child's skull was found buried beneath a stone slab covered in a series of strange markings; it was situated just a few metres away from the headless skeleton of a child – possibly the same individual. Not far from these graves the excavators found a group of low mounds; most were empty, but one contained the skeleton of a baby, and it is thought likely that the others were also originally used to house the remains of young infants, now completely decayed.

Further evidence of Neanderthal rituals involving dead children has emerged in a cave at Teshik Tash, in the Soviet Union, close to the Afghanistan border. Here the remains of a young Neanderthal boy have been found, his body laid out carefully amid ibex bones, with the horns of six ibexes neatly arranged in pairs to form a circle around his head. The ibex bones may be an indication that a supply of meat was buried along with the boy in order to sustain him during his journey to the next world. Again, the purpose and meaning of these rituals still remain shrouded in mystery; however, the combined evidence testifies to a growing fondness for ritual throughout Neanderthal times, which in turn suggests the evolution of an increasingly rich and more complex social life.

The most touching evidence of Neanderthal sensibilities was unearthed in 1957, in a large cave at Shanidar, in the mountains of northern Iraq. Here archaeologists found the skeleton of a Neanderthal man who had died some 45,000 years before. His bones told the story of a life that had been dogged by endless misfortune. His right arm was withered, probably as a result of an injury sustained during childhood. One of his eye sockets was smashed, indicating that he was blind in one eye, and as well as suffering from arthritis, he had sustained injuries to his legs, which must have made walking, let alone hunting, very difficult . Nevertheless, he not only survived all these handicaps, but carried on living until he was 40, a ripe old age for Neanderthals, the oldest of whom lived to only about 50. In the end, it took one final and spectacular piece of bad luck – a huge fall of rock inside his own cave – to crush the life out of him once and for all.

That the Old Man of Shanidar, as he is popularly known, managed to stay alive for as long as four decades is testimony to the care he must have received from his fellow Neanderthals. In a more primitive society, his evident incapacity to make an active contribution to the community would almost certainly have led to his being abandoned by the rest of his group. Seen as a liability, and a drain on the community's resources, he would have been if not thrown, then at least left to the wolves. However, it is clear that not only was the Old Man of Shanidar allowed to remain within

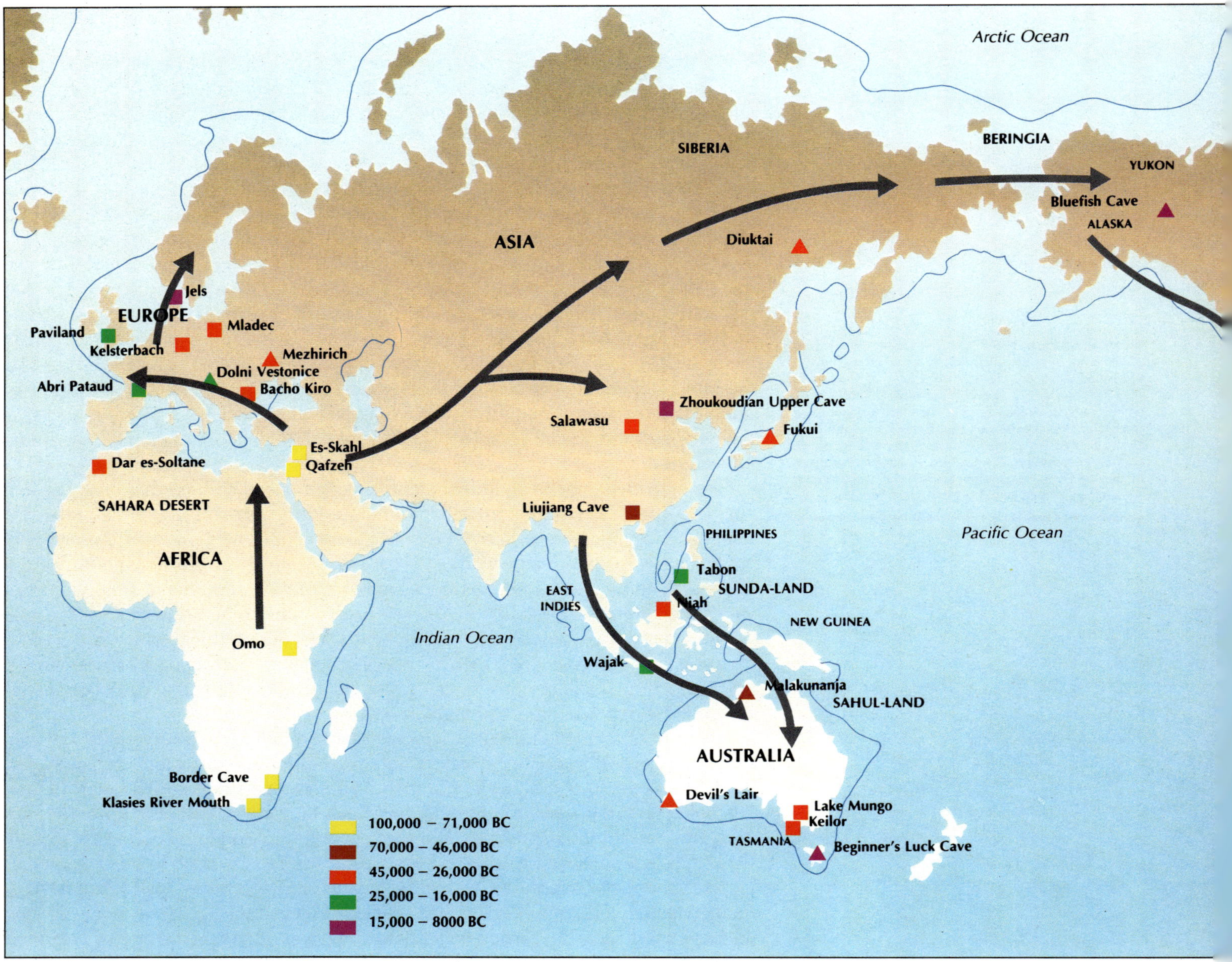

This map shows the routes along which it is conjectured that *Homo sapiens sapiens* spread out from his African homeland during the period 100,000 to 8000 BC. The earliest sites where his skeletal remains have been found are marked with a square; the sites of tools and other remains are marked with a triangle. The blue line indicates the extent of the world's coastlines at around 16,000 BC, when a fall in the sea level had created a land link between Asia and North America. The landmasses of New Guinea, Australia and Tasmania formed a single continent, known to modern geographers as Sahul-land. The Philippines, Malaysia and Indonesia were all joined on to mainland Asia, comprising the continent of Sunda-land.

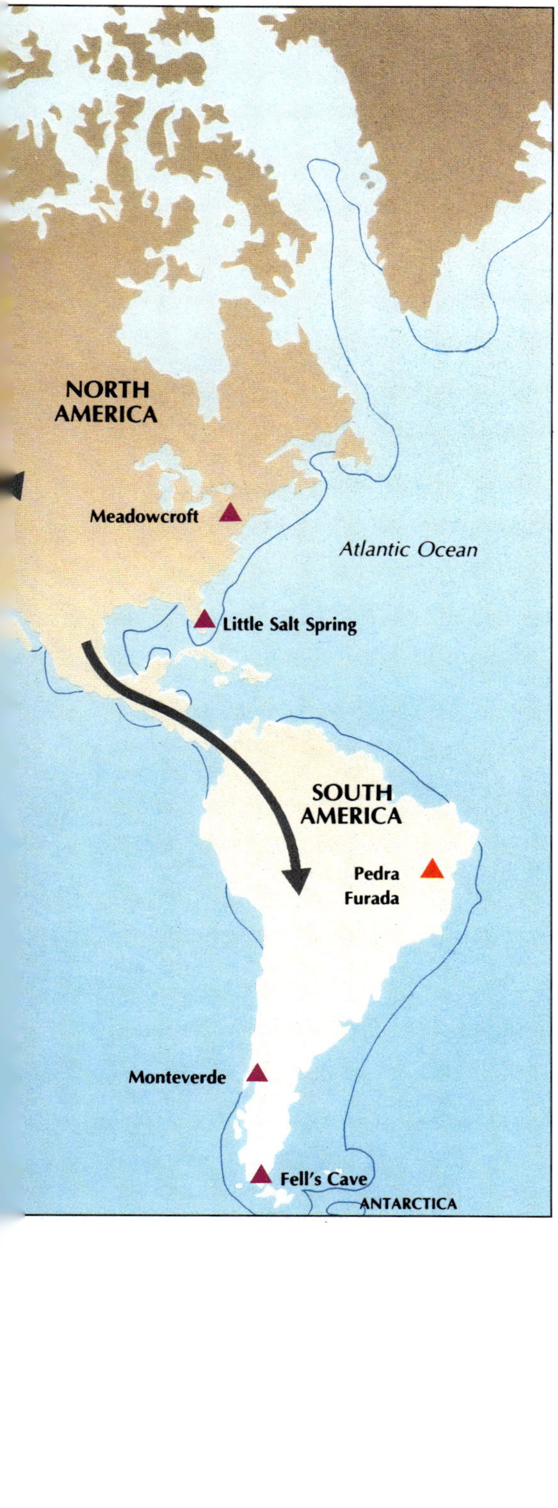

the community, but that he was also nursed back to relative health; all the wounds he sustained while alive show signs of having healed.

Nor did this caring come to an end when Neanderthals eventually died. Some of them, at least, were given a respectful burial. Another of the Neanderthals found at Shanidar, for instance, was surrounded by patches of flower pollen, which show that this individual was finally laid to rest upon a mattress made of woven pine-boughs and flowers, and that as a last token of esteem the burial party strewed the corpse with the brightly coloured blossoms of hollyhocks, grape hyacinths and cornflowers. It is possible that these flowers – which are used to this day in making Iraqi herbal remedies – were placed with the corpse as a medicinal aid in the afterlife; equally likely is the theory that they were scattered over the body in much the same spirit that moves modern-day mourners to lay flowers by a grave side.

Despite possessing these obvious sensibilities, however, the Neanderthals also seem to have had their limitations. For instance, excavations of the places where they lived have yielded no evidence for any kind of decoration, in the form of paintings or engravings, or for any personal adornment, such as jewellery. The practical applications of animal bone – for making tools – seem to have passed them by, and they seem also to have had little interest in travelling outside their own immediate environment; analysis of the stones from which they made their tools shows that they came only from rocks found in the immediate vicinity.

Indeed, all of the remarkable advances made by these earlier species fade into insignificance beside the explosion of change wrought by *Homo sapiens sapiens* – anatomically modern man. No more than 70,000 years after his first appearance in Africa (around 100,000 BC), *Homo sapiens sapiens* was to be found throughout virtually the whole of the habitable world. He had supplanted all earlier forms of human being and had colonized totally new areas such as the Americas, into which *Homo habilis* and *Homo erectus* had never penetrated. By 35,000 BC he had entirely colonized the Neanderthals' Eurasian territory.

The fact of his dominance is unquestioned. However, what remains unknown is the way in which he achieved it. Did this patently superior creature, with his glittering accomplishments in tool-making and hunting, arise in Africa in 100,000 BC and spread across the world squeezing out earlier types of hominid so that they became extinct? Or did *Homo sapiens sapiens* simply evolve naturally out of the existing populations of *Homo erectus* in Europe and Asia?

Conflicting answers to this question have been provided by investigations into one particular type of human DNA known as mitochondrial DNA, and found in cell particles whose function is to generate energy by burning fats and sugars. Of the two major groups of mitochondrial DNA carried by all human beings, the rarer type is found only in Africa. This much is agreed. However, where scientists differ is over the rate at which DNA changes over the centuries. Those who estimate one rate of change claim that the branching-off towards the non-African type occurred at some time between 180,000 and 90,000 BC, supporting the theory that a new species evolved in Africa and spread outwards, interbreeding with, or even supplanting, earlier Asian and European hominids.

However, this view is challenged by those who put forward a different estimate of DNA's rate of change, which would mean that the branching-off happened around 1.5 million BC – more than a million years before the emergence of *Homo sapiens*

sapiens in Africa. The colonizers cannot therefore have been *Homo sapiens sapiens,* who did not emerge until around 100,000 BC, but must have been earlier African hominids, namely *Homo erectus,* who during their expansion evolved first of all into archaic forms of *Homo sapiens* (including Neanderthals) and then, finally, into *Homo sapiens sapiens.*

Whatever the case, *Homo sapiens sapiens* was the only human who dominated wherever he went. Whole societies were swiftly and dramatically transformed by the innovations that he brought with him.

His first revolution was a technological one: *Homo sapiens sapiens* refined blade manufacture with astonishing results. Whereas one kilogram of flint had made *Homo erectus* between 11 and 45 centimetres of knife edge, and Neanderthal man some 200 centimetres, the *Homo sapiens sapiens* craftsman could produce a staggering 2,500 centimetres with the same amount. This was achieved not by hammering the flint core directly with a larger rock, but by interposing a soft chisel-point of bone or antler, with which he could coax off a large number of slices, each as sharp as a razor. This process would produce sturdy and elongated blades, but *Homo sapiens sapiens* was also able to manufacture smaller blades, measuring as little as two centimetres, which would be used for precision work. In addition to these knives, he made a variety of tools for special uses: scrapers for cleaning animal hides, and burnishers for smoothing them down and rubbing in colourings. He made saw-blades and awls, punches and borers, and he made flint burins, which were tiny points for the delicate task of engraving designs on antler and bone.

Rarities in Neanderthal times, these tools now became commonplace, as craftsmen invented the technique known as annealing. This was a process that involved making flint more workable by heating it up, perhaps in warm sand. The heat dried out the water within the flint, thereby making it more brittle and hence more likely to flake off in the direction the craftsman's chisel urged. Annealing would have been of use in the manufacture of the new, and deadlier flaked flint spearheads that *Homo sapiens sapiens* devised. Another of *Homo sapiens sapiens'* innovations was the joining of two different materials to produce composite tools, such as stone blades fixed to wooden handles by means of glue made from tree bark resin. The addition of handles gave the user a firmer grip and much greater leverage than if he had been holding the naked blade. These tools took a long time to make and were designed to provide many years of service.

Homo sapiens sapiens took another important technological leap when he began using animal by-products, in the form of bone and ivory, as the raw materials for tools and weapons. Less brittle than stone, both these materials could be formed into delicate, finely shaped points; sewing needles, for example, could be made out of bone, with animal sinew for thread. Bone also proved ideal for making barbed harpoons, used for spearing large fish and water mammals, and for fish-hooks and fish gorges – straight lengths of bone-splinter that were sharpened at each end so as to catch in the throats of biting fish.

Weaponry took a grim stride forwards. Not only did *Homo sapiens sapiens* craftsmen devise the bow and arrow, but they also came up with another device that was to increase the hunter's killing power. This was the spear-thrower, a short length of wood with a hook at one end that clipped on to the blunt end of the spear. Once the spear was in place, the hunter would balance the spear-thrower on his shoulder and then whip it sharply forwards, keeping hold of the spear-thrower by means of a thong handle, but dispatching the spear towards its target. Because of the centrifugal force generated by the thrower's wrist-snapping action, the spear would travel farther than one thrown by hand. Twentieth-century experiments have shown that a hunter using a spear-thrower can hurl a two-metre-long spear up to 135 metres, while a hunter without a spear-thrower is unable to manage much more than 65 metres. Nor did *Homo sapiens sapiens* confine his use of these new weapons to animals. A child skeleton from the Grimaldi cave, in northern Italy, was found with an arrowhead buried in its backbone; the world's first evidence of man committing violence against fellow man.

It may well have been the development of weapons like this, and of superior tactical ability, that led to the refinement of the *Homo sapiens sapiens* physique — he was only two thirds the weight of his predecessors. Neanderthals had had to be bulky. Without the benefit of the spear-thrower, they had been forced to tackle their prey at close quarters, using their spears more as stabbing tools than as missiles; in confrontations such as this, brute strength would have been essential for survival.

By contrast, the new hunters not only carried new weapons which allowed them to fight at a distance, but they had also devised new strategies, which enabled them to kill animals in far greater numbers than could have been achieved in direct combat. Game drives, for example, were now conducted on a scale hitherto unthinkable; whole herds of mammoths, horses, reindeer, ibexes and other large mammals were systematically driven over cliff tops to their deaths. *Homo sapiens sapiens* even took much of the labour out of searching for game. Instead of roaming across the landscape in the hope of chancing on suitable prey, he would simply set up his camp at some particularly narrow entrance to a valley and wait for the animals to pass by. Similarly, he would establish a base at river crossing points that he knew to be used each year by migrating herds; when his victims arrived, he would pick them off one by one from his vantage point on dry land as the animals, taken by surprise, floundered terrified in the water.

Hunter-gatherer groups now began to organize their lives more systematically than they had ever done before around the migrational patterns of mammals, the seasonal availability of fish or shellfish in sea or river, and the ripening of particular fruits or vegetables at various times and places. Annual timetables were evolved to accommodate all these key events, and the hunter-gatherers travelled around a fixed circuit of different camps throughout the year. They may still have been nomads, but there was nothing at all haphazard about their wanderings.

Characteristically, the members of a hunter-gatherer band would lay claim to a large area of land, each zone of which they would visit in turn. Since there was no telling when resources in one zone might suddenly run low, it was important to keep in routine touch with all parts of the territory, however far flung they were and however

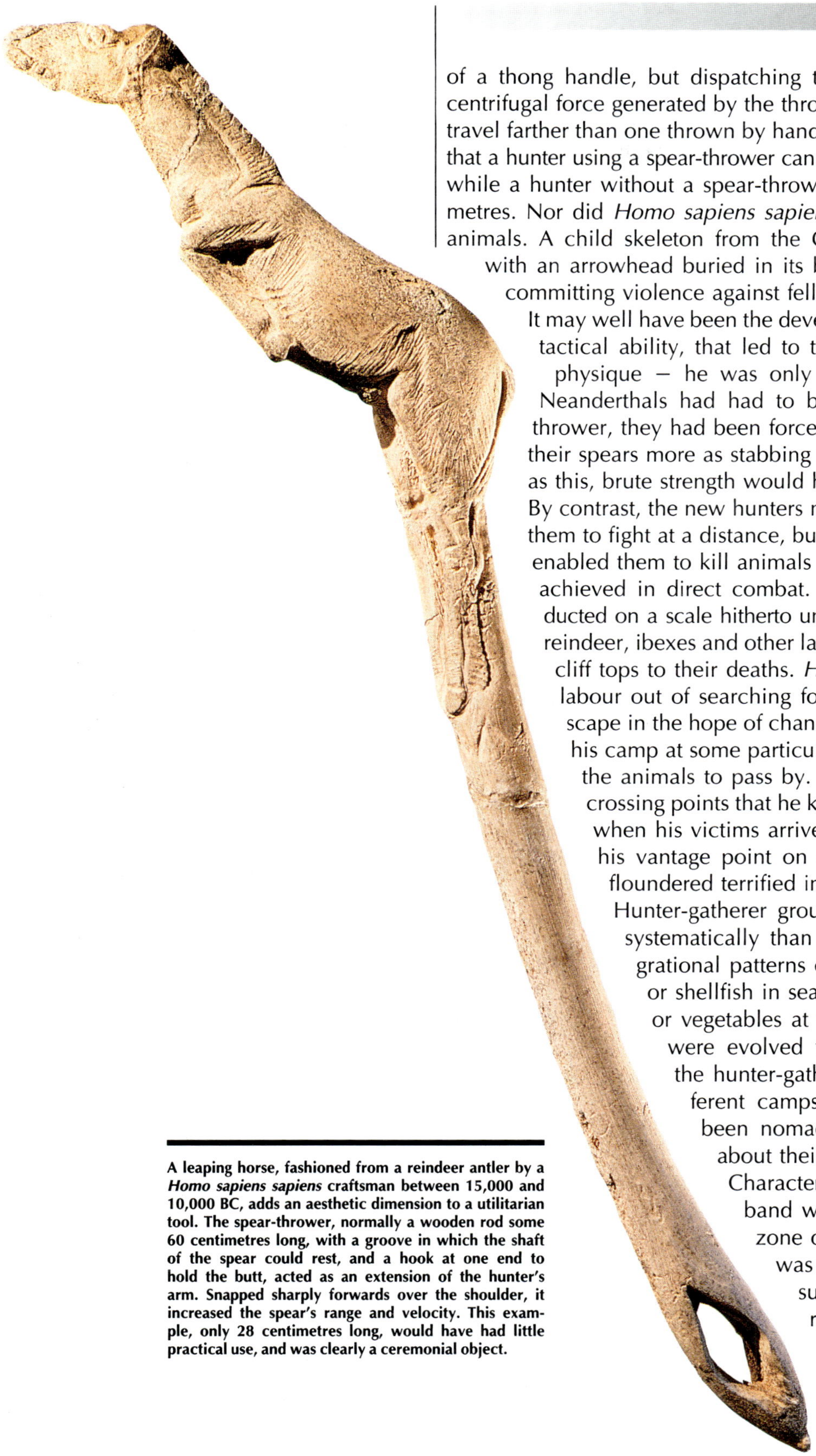

A leaping horse, fashioned from a reindeer antler by a *Homo sapiens sapiens* craftsman between 15,000 and 10,000 BC, adds an aesthetic dimension to a utilitarian tool. The spear-thrower, normally a wooden rod some 60 centimetres long, with a groove in which the shaft of the spear could rest, and a hook at one end to hold the butt, acted as an extension of the hunter's arm. Snapped sharply forwards over the shoulder, it increased the spear's range and velocity. This example, only 28 centimetres long, would have had little practical use, and was clearly a ceremonial object.

productive and comfortable one's immediate surroundings might be. At the same time, even though they were based in different places at different times of year, these hunter-gatherers developed a very strong sense of home. Whereas non-human primates had always stayed together when searching for food, these humans probably divided their labour, some of them going off on hunting expeditions that could last for several days, while others searched more locally for nuts and vegetables. This constant scattering of the group made it essential that they all had a base to return to, and this need for a secure focal point around which these activities could revolve, gave the camps established by *Homo sapiens sapiens* a feeling of semipermanence that was lacking in those of their predecessors. Whereas excavations of earlier hominid sites have uncovered signs of domesticity amounting to little more than ash from constantly tended fires, investigations of known *Homo sapiens sapiens* bases have produced evidence of a far greater attention to home activity and comfort.

They made their huts out of stone, wood and animal bones and hides. Inside, hearths were set over small trenches that led out into the open; this ensured a constant supply of air to the flames, the draught acting like bellows in keeping the fire going.

The search for warmth, in fact, was something that preoccupied *Homo sapiens sapiens* for much of the time in northern latitudes. Born in the tropical regions of Africa he spread into zones that were colder than his physical evolution had prepared him for. It is likely, although no samples of skin or hair survive to corroborate this, that *Homo sapiens sapiens* at first had the dark complexion and relatively hairless body that make life more comfortable in a hot climate, but are not as suited to life in the chill regions of central Europe; lighter skins were only a gradual, later development. Shivering *Homo sapiens sapiens* found ready-made homes – in

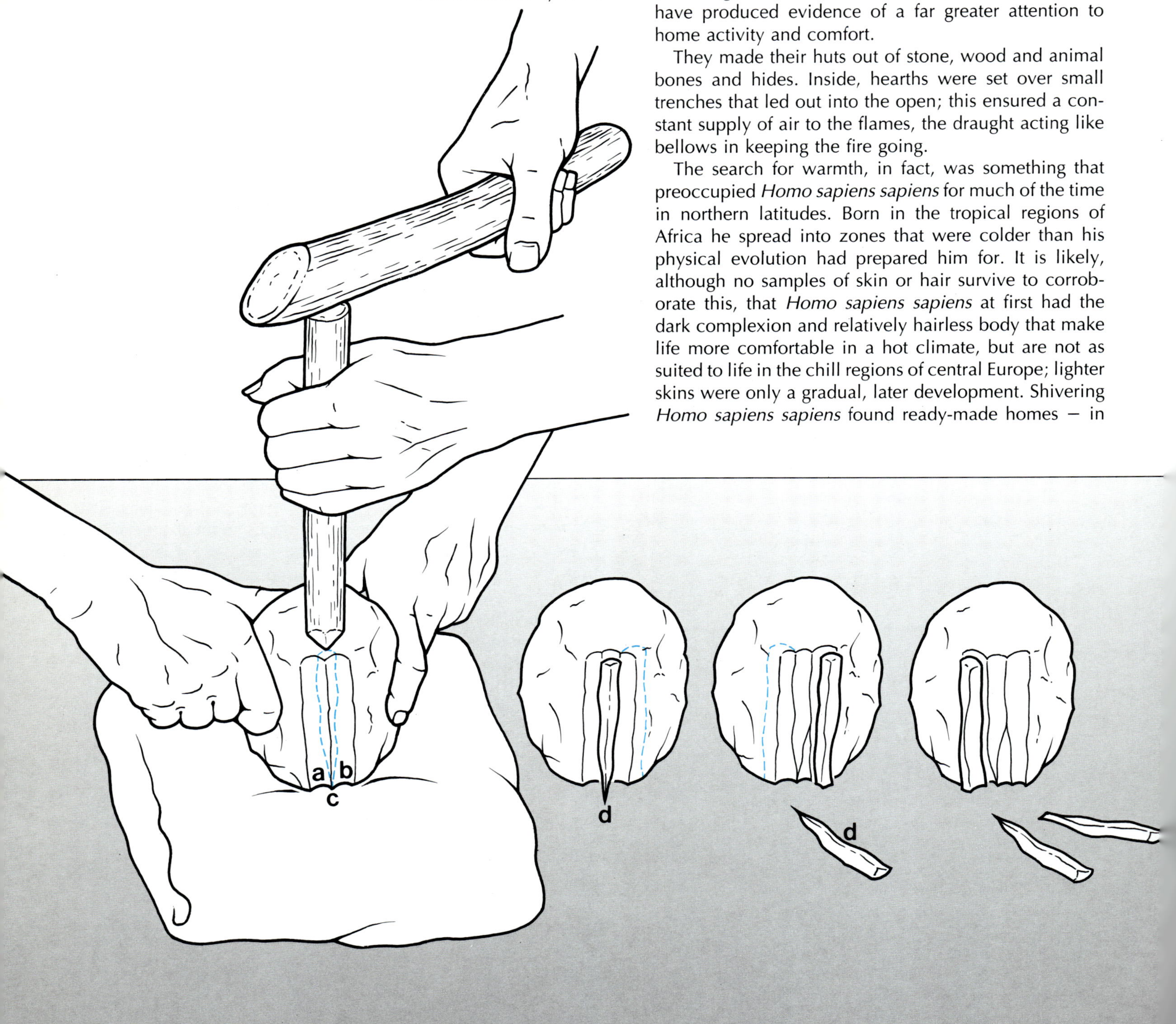

the form of natural caves — only in relatively few places, such as those discovered in southwestern France. Elsewhere, therefore, like his predecessors at Terra Amata, he had to devote a large proportion of his time and energy to providing himself with shelter from the elements.

Eventually, this tendency towards leading a more settled life would result in the evolution of social patterns that are common to all static societies. The beginnings of hierarchy and of rank, and of the differences between the rich and the poor, would all have begun slowly to manifest themselves. For the time being, though, the values of hunter-gatherer society were retained in more or less pure form. Living in such small groups, so vulnerable to famine or to attack from predators, and so dependent upon each others' labour, hunter-gatherer societies would have tended, then as now, to be highly egalitarian. Such bands had no headmen or chiefs; instead, like their modern equivalents — such as the !Kung people of the Kalahari Desert — they employed a system of "dispersed leadership", which recognized the particular talents and experience of each participating individual. Decisions were taken collectively, without formality. There was no strong sense of privacy, so any grievances were quickly made public and settled without being allowed to fester. The notion of private property, too, was alien to the hunter-gatherer. Instead, the importance of sharing was deeply ingrained: elaborate rules existed for the division of meat and other produce, dependent upon who had killed or gathered it, and to whom they were related or obligated. Again, if we can go by our observations of hunter-gatherer groups today, teamwork must have been of the essence. The individual generally would have disliked standing out among his or her companions, and modesty about one's achievements was a much-prized virtue — successful hunters would often go to extravagant lengths to play down their particular skills or triumphs, however impressive they may have been.

In warmer climates, where most modern hunter-gatherers live today, the plants and vegetables gathered by the women provide the staple diet, the meat killed by the men constituting no more than 40 per cent. For early *Homo sapiens sapiens* this must have been the case, too, although in colder regions the gathering and storing process was far more rushed and labour intensive. In Europe, for example, the annual picking

Simple mass-production methods enabled *Homo sapiens sapiens* to make tools more efficiently. The drawings on the opposite page show how a number of blades were cut from a prepared core of flint: by hammering a chisel of bone, antler or wood, the craftsman first gouged out two channels (a and b) from the flint, thus creating a pointed ridge (c). He then positioned the chisel at the top of the ridge and hammered downwards, splitting off a sharp sliver (d) and in the process creating two more ridges. He would continue in this way until he had used the whole flint. The slivers were then made into spearheads or knives, or used to fashion more sophisticated products such as the antler harpoon-head *(right)*.

THE TOOL-MAKER'S CRAFT

Barbed harpoon-head carved in France, 12,000 BC.

season was a matter of weeks rather than months; furthermore, staple foods such as beechnuts and hazelnuts had all to be painstakingly shelled by hand, and provided only small nutritional returns in the form of the kernel at the centre. It is clear, then, that the contribution of those doing the gathering was of crucial importance. Hunting expeditions depend a great deal upon luck; in the animal world, for instance, wolves manage only one kill for every 10 attempts and even lions fail more often than they succeed. By contrast, the collection of vegetable foodstuffs was predictable in its outcome; a woman knew that one day's foraging would provide her with three days' food for her family.

This being the case, it was vital for the survival of the group that the women were not prevented from their regular gathering duties by having to endure constant rounds of childbirth. But how to prevent conception? The solution to this dilemma was a simple and probably unconscious form of birth control, based on the principle that mothers who are breast-feeding their babies have a low incidence of conception; this is because the baby's continual sucking action stimulates the production of prolactin, a hormone that can inhibit ovulation. In contemporary hunter-gatherer societies the women breast-feed their children until they are at least three years old, thus enabling them to maintain a three to four-year gap between children. Earlier societies will probably have evolved a much similar pattern, since bearing children any more often than this makes the twin burdens of child care and foraging work impossible to sustain with any degree of efficiency, while bearing children less often might endanger the continuation of the group.

As well as operating this rudimentary, though effective form of family planning, hunter-gatherers must have followed certain patterns of self-regulation for the good of their society as a whole. In order to provide themselves with maximum security and optimum efficiency in finding food, hunter-gatherers today form little bands of between 25 and 30 people, each of the groups including around 12 adults and their children. Approximately 20 such groups will be loosely affiliated as a tribe, which therefore comprises some 475 to 500 individuals in all. This figure is just large enough to prevent members of the group from having to look either outside the tribe or among their close relatives for sexual partners, practices that can undermine the self-sufficiency of a tribe, or create through inbreeding genetic defects that could make the group vulnerable. Regular tribal gatherings – ostensibly for combined hunting or foraging operations – would have provided opportunities for socializing, and for meeting possible partners from other groups.

This pattern first established itself during the early days of *Homo sapiens sapiens*, as a way of organizing the world's population at a time when it was growing extremely quickly. The welfare of the group was always paramount.

Such an emphasis on cooperation and sharing made communication between members of the group indispensable. While many mammals and birds have distinctive alarm calls that they use to warn their companions of dangers, and while some primates even have several different calls for indicating different sorts of danger, no other species has anything that remotely approaches the linguistic gifts of humans. Studies of the skulls of early hominids have allowed specialists to reconstruct the vocal tracts of australopithecines and *Homo erectus*. They have discovered that the australopithecines – like modern-day babies – had flat palates, which would have left them unable to vocalize many more sounds than can apes today. The palate of *Homo erectus* was a little more curved, allowing him a slightly greater range, though this was probably not much more than that of a seven-year-old child today. As for the Neanderthals, it is not known for certain whether they were capable of truly articulate speech. Most of the specimens that have been found indicate that the Neanderthal larynx, containing the vocal chords, sat much higher up in the throat than in human beings today. This in turn restricted the size of the pharynx – the area above the larynx in which the speaker, by contracting the muscles in his throat, modulates the sounds produced by the vocal chords. This therefore reduced the number of vowels and consonants that Neanderthals were able to utter. At the same time, however, Neanderthal remains displaying a throat structure that is essentially modern have been discovered in the Middle East.

On balance, it seems unlikely that the majority of Neanderthals had any great degree of vocal ability; no examples have ever been discovered of what we might call more refined Neanderthal endeavours, in which speech and the communication of ideas would have played a part.

It was only with the arrival of *Homo sapiens sapiens* that the palate became fully curved, and that language began to attain any great variety and expression. Two new developments made this possible. First of these was a gradual increase in the capacity of the human brain, and with it that part of the brain which controls the translation of thought into speech. And second was the growing number of cultural and social demands that were being imposed by the increasingly complex nature of the communal society in which *Homo sapiens sapiens* lived. When the earliest humans first walked upon the planet, the actions of the individual, however ingenious, had not demanded the extensive use of language. None of early *Homo sapiens sapiens*' breakthroughs in tool-making, for instance, had been dependent on the ability to speak. Neither had his developments in hunting technique. Hunter-gatherers tend to do their hunting in silence, and in any case other animals – African wild dogs, for instance – are capable of carrying out highly elaborate group manoeuvres without the benefit of language. Linguistic competence did not become a necessary attribute until humans started to live in settled and organized groups, where they evolved systems of planning, of allocating tasks, and of sharing meat and other produce – systems that went beyond the immediate and self-evident needs of the moment to cater for what was not yet apparent, not directly obvious: for what, in short, needed to be explained.

As humans' linguistic ability developed, so it took them further and further into the realm of the abstract and the symbolic, opening up a world that extended beyond the here and now. Previously, artistic activity seems to have been limited to the adornment of practical objects. Suddenly, however, around 30,000 BC, there came a great outpouring of creativity across the populated world. An astonishingly varied range of portable art works – among them strange clay and stone figurines in human and animal forms, bone and ivory pieces engraved with dazzling virtuosity, delicate beads, shell necklaces and other ornaments – and more large-scale and permanent items such as the awesome cave paintings discovered in France and Spain, provide eloquent testimony to this new, artistic sensibility. The precise purpose of these works remains a mystery, and it has yet to be established whether the motivation behind such images was religious, educational or aesthetic. One of the few conclusions we can draw with any degree of certainty is that most early human beings were right-handed. An analysis of the hand stencils that have been painted on cave walls shows that the majority of hands depicted are left hands, which would suggest that the artists used their right hands to apply the paint; at the French cave of Gargas, for example, there are 136 left hands painted on the wall, as against 22 right. But even though the cave paintings pose many more questions than they answer, there can be no doubt that they and other examples of early decorative art indicate a human race that was more than ever before in control of the immediate necessities of his physical existence, and therefore with the leisure and confidence to begin developing a richer cultural life for himself.

Homo sapiens sapiens' advanced system of social organization helped to equip him for the enormously diverse range of environments that he encountered as he expanded gradually outwards from his African source, up through the Sahara and North Africa to central Europe, and then right across Asia to Siberia and the Americas in the north, and to Indonesia and Australia in the south. The impetus behind this expansion was the constant search for richer and more abundant food supplies – a search that was to continue until all the regions of the Old World, and ultimately those of the New, had been populated by humans.

That they had succeeded in colonizing the globe was due to their capacity for teamwork. Their highly developed, deeply ingrained systems of cooperation and interdependence made them far more flexible than earlier hominids had been, and thus able to make the most of whatever situation they found themselves in. A look at how some of these earliest modern humans lived, in sharply contrasting settings, illustrates this remarkable adaptability.

On the coast of southern Africa, in a cluster of caves and crude rock shelters at the mouth of the Klasies River, lived some of the very first communities of *Homo sapiens sapiens*, long before the species appeared in Eurasia. Between 100,000 and 70,000 BC, as sea levels fluctuated through glacial cycles, the site remained close to the

coast; after that, the waters receded, ultimately to a distance of around 65 kilometres, and the complex was abandoned.

However, for those 30,000 years, successive generations of Klasies River Mouth people enjoyed a favoured existence. Throughout this period the region enjoyed mild climatic conditions and furnished abundant food supplies both from the land and the sea. Parties of women would forage in the bush, while smaller groups of men — two or three at a time — would make longer journeys inland, hunting the large mammals on which the entire community could feast: blue antelope, Cape buffalo, black wildebeest and the now extinct giant buffalo, which might weigh anything up to 1,600 kilograms. Some species seem to have been avoided — no remains have yet been discovered at the site of the choleric bushpig, for example, which was always liable to turn ferociously upon its attacker. It was probably regarded as being too dangerous to be worth hunting.

As well as satisfying immediate food needs, the meat from these kills would also feed the Klasies River Mouth people during the winter, when meat was scarce; any leftovers would be dried in the smoke from their fires, a simple but effective way of preserving meat in a warm climate. The cave-dwellers also used fire to improve their yield of vegetable crops; this involved burning patches of the bush in order to stimulate the growth of smaller food plants, which otherwise might have been choked by larger, inedible vegetation.

The people of the Klasies River Mouth did not look only inland for their provisions. The shoreline's myriad rock pools were systematically scoured at each low tide for limpets, periwinkles and mussels. Sea birds were killed for food, and there is evidence that whales and small dolphins were also consumed, probably after being washed ashore. Cape fur seals, spending the winter on nearby beaches, were often slaughtered, though this was probably as much for their warm fur as for their meat.

These drawings reconstruct a dwelling of the type that was built by *Homo sapiens sapiens* hunter-gatherers on the treeless steppes of the Ukraine in around 13,000 BC. Around five metres in diameter, the framework was built entirely of mammoth bones, weighing some 21,000 kilograms *(opposite page)*; arching tusks formed the doorway, and each component was held in place by pins made from smaller bones. A covering of turf reinforced by loosely sewn furs and skins, with further layers added according to weather conditions, turned this small, domelike structure into a cosy home through the long, harsh winter months.

The sea, therefore, proved an important source of nourishment, in particular because it supplied much-needed sustenance during the long winter months, when edible plants were harder to find. The addition of seafood to the basic hunter-gatherer diet transformed this community's position at a stroke, giving them an extra layer of security against food shortages.

Though the settlement at the Klasies River Mouth fell into disuse around 70,000 BC, the people who inhabited a settlement just down the coast at Nelson Bay – some 60,000 years later – had clearly absorbed many of the lessons that had been learnt by their forebears. They too lived in and around a broad-mouthed cave, with fresh water conveniently trickling from a spring at the back; like the Klasies River Mouth people, they also went on hunting expeditions inland, although they now felt able to tackle the fierce bushpigs and wart hogs, and had added ostriches and baboons to their diet. Technologically, though, they had advanced well beyond their predecessors: they made more sophisticated tools, they built stone rims for their hearths and they erected a windbreak of wooden posts draped with hides or brushwood, to protect the entrance to their cave.

The people of Nelson Bay had come up with a number of efficient methods of fishing, most notably with the fish gorge. This small, sharpened sliver of bone or wood, measuring perhaps five centimetres in length, was secured to a line made either of leather or of animal sinew, and lodged sideways in the fish's throat when swallowed, allowing the fisherman to haul in his catch. The gorge was superseded, however, by the fishing net, also thought to have originated in South Africa. This was woven from vegetable fibre or hide thong and weighted with specially prepared cylindrical stones. The arrival of the net marked a crucial breakthrough, in that it made large-scale harvesting of fish possible for the first time. The sea thus became a reliable and regular source of food, allowing coastal communities to stay in one place for the greater part of the year.

Meanwhile, 6,000 kilometres to the north, several separate communities had, from around 15,000 BC, been developing a very different way of life. They lived on the Kom Ombo plain, a 400-square-kilometre stretch of land situated on the eastern bank of the River Nile, in what today is southern Egypt. An area of boggy pasture for much of the year, the plain offered a plentiful supply of lush reeds and edible grasses that were much to the liking of the large herds of wild cattle that inhabited it. During the hot, dry months between March and August, before the annual rains began, the cattle would be joined by antelopes and gazelles, which were drawn to the plain in search of food and water. It was the presence of these animals that initially attracted hunter-gatherers to Kom Ombo; once there, they discovered that the plain was teeming with many other creatures, especially between August and October, when the rains in East Africa built up a large head of water upcountry and sent wide river channels rushing down across the plain. As well as the wallowing hippopotamus, a formidable source of meat when it could be overcome, the hunter-gatherers found on the plain a rich variety of edible water birds, both resident species and migrants passing through. Among the many diverse types that they could choose from were ducks, geese, cormorants and even herons and eagles. Meanwhile, beneath the surface of the water, shoals of catfish, perch and other freshwater fish abounded, along with the soft-shelled turtles of the Nile.

Such a wealth of natural riches allowed several hundred people to occupy this relatively small area at the same time – a remarkable population density given the many hundreds of square kilometres needed to support a single such group in less accommodating environments. Following traditional hunter-gatherer patterns, the people of Kom Ombo lived in groups numbering between 40 and 50 people. Each of these groups retained its own specialized way of life, which was signalled by its own slightly different tools and techniques of manufacture. Some of these bands had become so settled by this stage that they even practised a very basic form of agriculture, gathering their supplies of grain on an increasingly systematic basis from the cereal grasses that grew all over the plain.

Altogether more hostile conditions prevailed on the steppes of central Europe and Asia, where long, bitter winters brought temperatures of 40 degrees Celsius below freezing, and short-lived summers never got warmer than 20 degrees. Even here, however, *Homo sapiens sapiens* survived and even thrived.

The steppe held obvious appeal for any hunter-gatherer band: herds of reindeer, bison, horses, mammoths and woolly rhinoceroses. But rich as the steppe was in wildlife, it presented a decidedly unfriendly exterior in other respects. Trees and plants grew only sparsely, and without any significant hills or an area of forest cover to provide natural screens, the biting winds tore across the plains unhindered. Finding adequate shelter was of paramount importance in such adverse conditions, but there were few caves to provide ready-made protection from the elements, so human beings had no choice but to construct their own housing, using whatever materials came readily to hand. In parts of eastern Europe where wood was scarce, mammoth bones were a commonly used substitute.

Indeed, for the community which around 25,000 BC settled semipermanently at Dolni Vestonice, in southern Czechoslovakia, life seems to have revolved almost entirely around the mammoth. From the large, artificially shaped hollows that now indicate where the settlement's four huts once stood, come finds that demonstrate the great beast's importance to the members of this community. As well as furnishing them with a substantial quantity of meat, it seems that the mammoth hides were scraped down and used to cover the huts, and mammoth bones to hold those hides in place. Mammoth tusks and bones also provided the framework for a defensive stockade around the huts, and the settlers overcame the shortage of wood fuel by burning mammoth bones instead. Although it was initially difficult to light, the bone gave off a hot flame once it was burning, and must have made a cheering warmth inside the largest of the four huts, a 92-square-metre construction in which as many as five hearth fires might be alight at one time. It was while sitting around these hearths that the people of Dolni Vestonice fashioned little models of the lionesses, rhinoceroses and other animals that they encountered out on the steppe; some were made of clay, others – inevitably – were carved out of ivory mammoth tusk.

It had been as a base for hunting the mammoth that the Dolni Vestonice site had been established in the first place. The site chosen by these hunter-gatherers for their village must have lain very close to one of the chief routes used by the mammoth herds on their winter migration southwards. Each year, as the great beasts moved down the valleys of the Dyje, Morava and Danube rivers, the hunters were lying in wait for them. There was little noble or stirring about the mammoth hunt; no epic struggle for mastery between man and beast. Instead, every autumn, year in and year out, the hunters ambushed the herds at the same spot, raining arrows and spears upon the terrified animals and then butchering them on a near industrial scale. While the hunters took as much meat as they could from their kills, they inevitably left large amounts behind, because the massive carcasses were too heavy to move. The meat the hunters did manage to cut free would be dried in the wind and sun on a nearby hillside or else stored in frozen pits.

As well as killing mammoths, the people of Dolni Vestonice got meat by scavenging – that oldest of hominid strategies – and by hunting down the other large mammals that grazed on the steppe, including reindeer, horses and bison. From the carcasses of these animals they derived not only meat to eat but also hides with which to cover both huts and human beings.

The clothing of the Eurasian hunter-gatherers most probably resembled that worn by Eskimos today: hooded tunics made of hide or fur, with tightly sewn seams to trap the heat, and leggings that could be tucked snugly into hide boots. Fur socks and mittens gave the feet and hands further protection against the biting cold. The fur came from hares, foxes and martens, which the hunter-gatherers caught in rudimentary snares; these consisted of rawhide nooses held taut from bent saplings, which sprang up when disturbed by a passing animal, jerking their victims into the air and leaving them dangling there until frostbite or returning hunters applied the *coup de grâce*. Birds too were hunted, caught in traps on the ground, or else brought down in mid-flight with arrows or other missiles.

In spring and summer, the people of Dolni Vestonice had to compete for their prey on equal terms with other carnivorous animals. However, during winter, humans' mastery of fire gave them a clear edge over their rivals, since it enabled them to thaw out the carcasses of animals that had died in the snow, rather than having to wait until nature did the job for them, in spring.

Not that they would starve if no meat was forthcoming. Like other communities of this region at the time, the people of Dolni Vestonice probably supplemented their meat diet with fish from surrounding rivers: pike and perch were both popular, but the salmon quickly became the staple fish, once its habits were understood. Like the mammoth, it was rendered vulnerable by the predictability of its annual migration. Salmon swimming upstream to spawn had each year to run the gauntlet of predatory men, who flocked to the river banks at the appointed time to catch the migrants. Often they blocked the rivers with weirs built of stones, channelling the fish into confined spaces where they could easily be speared, perhaps with the leister, a Viking word meaning "stabbing tool". This was an ingenious three-pointed spear, the central tine of which impaled the fish, while the two curved prongs on either side ensured that it did not wriggle free.

The salmon-hunting season lasted only a short time. It took fewer than three weeks for all the fish to pass by a particular point, and some 40 per cent of them would swim past within the space of just three days. It was vital to make the best of this, the most critical time, and so, while the men fished feverishly against the clock, the women would frantically gut, dry and store the catch. Theoretically, the salmon run should have been a major source of food; however, the sheer scale of the operation generally outweighed the physical resources of a small hunter-gatherer band, who would be unable to grasp hold of more than a small proportion of these showered riches. This meant that salmon, though not in short supply, could never amount to more than just a supplement to their main meat diet.

For the inhabitants of Dolni Vestonice, food remained of necessity the first priority. However, they did not neglect the finer things in life. There was certainly music,

MARKS WITH MEANING

The possibility that humans may have devised a system of recording information as early as 30,000 BC is suggested by the marks on a section of reindeer bone recovered from a site in the Dordogne, France. The whole bone plaque is shown actual size at the bottom of the picture; the enlargement above it is overlaid with the outlines of the 69 engraved marks.

Analyzing these impressions, the U.S. researcher Alexander Marshack concluded that far from being random doodles, they were made by design, possibly with a number of different tools and over an extended period of time. Relating the marks to the waxing and waning of the moon, Marshack proposed that they comprised a consecutive record of lunar phases.

The looping line on the left, corresponding to the arrangement of the marks on the bone, shows the order in which the marks may have been intended to be read.

whether as an accompaniment to religious ritual or as entertainment in its own right; somebody in the community played upon the simple flute, made from a section of hollowed-out bird bone, that was unearthed at the site many thousands of years later. This musician was perhaps accompanied by others, playing some of the mammoth-bone percussion instruments of the time: the xylophone made of hipbones, drums made from shoulder blades or hollow skulls, and rattles made of jawbones. The visual arts were also catered for, although the vast expanses of blank cave walls of France and Spain were not available to the Czechoslovak artists of the time. Instead, works of art were formed in terracotta. A fifth hut at Dolni Vestonice, standing apart from the others outside the stockade, housed the world's first known pottery kiln — an oven enclosed within earthen walls, in which pots and sculptures of soft clay mixed with powdered bone could be fired into more lasting forms. These first potters sculpted animal heads — foxes, bears and lions — and they fashioned strange little statuettes of female figures to whom they gave exaggeratedly swelling breasts and buttocks. Fascinating as these objects are aesthetically, they carry still greater significance in terms of the technical advance they represent: the potters' clay used to make these pieces is the earliest example of two different substances being compounded to produce a useful third material.

Inevitably, the success of *Homo sapiens sapiens* in adapting to such a wide range of environments led to a boom in population through the closing millenniums of the ice age. As any area threatened to become crowded and its resources over-exploited, whole hunter-gatherer bands — or smaller splinter groups, perhaps of young people seeking greater independence — would move out in search of virgin territory. Little by little then, the earliest modern humans spread out across the whole of the Old World, acquiring new skills and customs as they went.

It was in just such a way that Australia was first colonized, around 50,000 BC, by peoples who had gradually moved southwards in search of more fertile pastures over many thousands of years. On their way they progressed through what is now the island region of the East Indies, but what for long periods of time then was a continuous expanse of dry land. It is not known for certain whether the route these people took led them through southern China and the Philippines, or from Southeast Asia to Indonesia. What we do know, however, is that the next stage in their journey must have been a long sea voyage of at least 90 kilometres across the Java Trench, the deep expanse of water that stood between these migrating peoples and the neighbouring continent of "Sahul-land", which up until 8000 BC joined Australia, New Guinea and Tasmania in one great landmass.

It remains a mystery as to what sort of vessels these early mariners used for their voyage. Certainly, more recent Australian Aborigines have had no boats equal to the task, which leads to the theory that bamboo — a plant native to Southeast Asia but not Australia — was used in their construction. But whatever the composition of their boats, and however flimsy or makeshift they might have been, it seems clear that the purpose of these first Australians was both clear and well-formed. They were seeking a new life in a new land, and were prepared to set out into the unknown, across what might, for all they knew, have been an endless tract of ocean, in order to find it. Mathematics reinforces this theory: had one lone raft arrived accidentally — due to navigational error or being blown off course — its passengers would not have been able to parent children in sufficient numbers for the subsequent population to reach the size it did. It seems certain, then, that deliberate migrations of several rafts, probably staged in several phases, took place.

Initially, the settlers' existence remained based on the coast; the sandy beaches and mangrove swamps that they discovered in northern Australia were very similar to those of Indonesia's low-lying islands, and the animals that they found there — crocodiles, turtles, sea birds, fish and shellfish — as well as the fruit trees and other vegetation, also closely resembled those their ancestors must have hunted and gathered. The new settlers maintained this life style as they worked their way around the coast; it was the way they knew best, indeed the only way they knew, and in any case the Australian interior seemed remarkably unenticing, with its giant marsupials and its strange flightless birds.

However, having reached the south coast of the continent, some bands gradually began working their way inland, following the course of the great Murray-Darling river system, and they established a freshwater version of their old coastal life around the shores of Lake Mungo. Here, for nearly 20,000 years, the lake-dwellers lived a comfortable and settled existence, catching the carp, cod and crayfish that abounded in the lake, and gathering freshwater mussels. Brief sorties into the interior brought back a variety of food in the form of emu eggs, wombats, wallabies and rat kangaroos — small, rat-sized hopping marsupials. But around 16,000 BC, the climate in Australia grew increasingly dry, culminating in a drought, which dried up Lake Mungo. Like the other bands of Aborigines that were spread out across the continent, the erstwhile lake-dwellers now had to adapt their life style to a desert climate.

Faintly discernible amid a riot of lines engraved on a stone plaque, what is clearly a horse's head strains against a crude halter *(left)*; the simplified outline is shown in the diagram below. Excavated at La Marche in France, this stone image would appear to indicate that horses were being ridden or led in harness as early as 12,000 BC. There is also evidence to suggest that ancestors of the domestic dog may have been trained to stand guard against invaders over hunter-gatherer settlements.

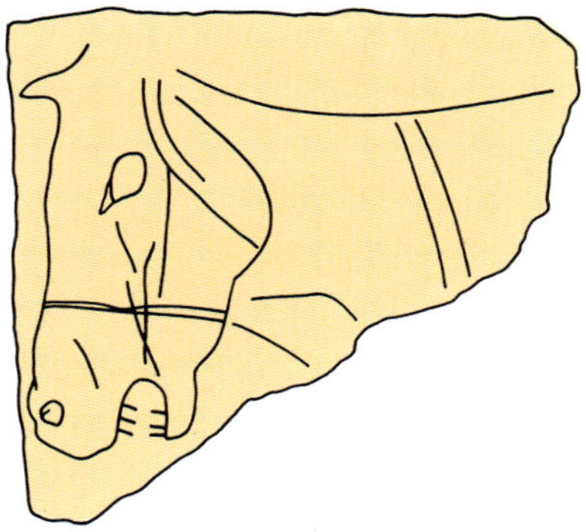

Stricken with perpetual drought, vast tracts of land stretched out featureless and bare, save for dried, twisted trees and occasional patches of parched scrub. Faced with this forbidding environment, the inland Aborigines soon established a life style based around the hunting of kangaroos and other marsupials, along with the harvesting of wild millet seed. It was not long after this that the giant marsupials became extinct, probably because bush burning by the Aborigines, intended to stimulate the growth of their own food plants, destroyed the shrubs on which the animals fed. When, in 8000 BC, the end of the ice age resulted in a flooding of the more hospitable coastal settlements, forcing back the inhabitants into the arid zones, it marked the beginning of an inland, desert-based way of life that would last for many thousands of years. Today, the memory of any earlier coastal existence has long since faded among the Aborigines who inhabit the interior.

Compared with this epic story of a people's setting out across the sea in search of a new continent, the arrival of *Homo sapiens sapiens* in North America seems to have been a rather low-key event. There are two main theories of how human beings first arrived in the Americas. According to the traditional version, hunter-gatherer bands who had settled on Beringia, the icy stretch of land which from 25,000 BC joined Asia and America, simply moved eastwards in search of fresh resources. Had they not already made this move, they would have been forced to do so towards the end of the glacial period, around 13,000 BC, when the sea level rose and once again the waters closed over the top of Beringia.

Once in America, these new immigrants found little in their new surroundings to surprise them: animals too had made their way across the land bridge from Siberia, and a variety of plant seeds had also been blown over, both recently and in the more distant past, when previous glaciation cycles had thrown the route open. For these first settlers, therefore, life went on much as before, except that to the south there now stretched out before them the whole of the Americas, a land of almost limitless opportunities for hunter-gatherer peoples.

For a long time, however, the path leading to those opportunities was blocked; beyond Alaska and the Yukon territory two enormous ice sheets almost met to erect a barrier that was well-nigh impregnable. One, known today as the Laurentide, inched southwards and westwards across northern Canada and America; gouging and scraping its slow, laborious way across the landscape, it had joined its counterpart, the Cordilleran, whose frozen heart lay in the centre of the northern Rockies. Although some brave souls must have made the hazardous journey through the narrow, ice-free corridor that separated the two sheets, people would not have considered migrating southwards in any significant numbers until the ice sheet began to retreat, which took place around 10,000 BC.

Once the way was clear, though, *Homo sapiens sapiens* flooded south, attracted by the astonishing richness of game the virgin continent provided. As usual, the hunter-gatherer bands kept moving onwards in search of new resources, constantly occupying new territories. They were rarely disappointed: these were times of extraordinary plenty. The mammoth, the bison and the other large grazers that teemed in the heartlands of North America, unused to the ways of man, were far more easily hunted than their relations on the steppes. Game drives proved devastatingly effective in these circumstances, and bison and other large mammals were stampeded over cliffs in their thousands, much of the meat that resulted simply being left to rot.

In this bonanza several species were quickly hunted to extinction, including the mammoth, the ground sloth and the elephant-like mastodon.

Meanwhile, the human population, its numbers having been boosted as a result of such prosperity, continued the slaughter of these animals on an even larger scale. The food resources in each new territory started to run low rather quickly, and the search for new hunting grounds soon took the first humans through to the Central American isthmus, and on to South America. By 10,000 BC the first hunter-gatherer bands had reached as far as the southern tip of Chile, and the occupation of the Americas from top to bottom was complete.

This traditional theory of when settlement first took place in America insists upon a surprisingly late date of arrival, and a remarkably swift rate of expansion through the Americas after that – some 18,000 kilometres in 1,500 years. Supporters of the most popular rival theory claim that this version of events sits uneasily with the good deal of evidence of human activity in South America that seems to suggest *Homo sapiens sapiens* arrived on the continent many thousands of years earlier. Previously, a precise date had proved difficult to establish, since no sign of any human life before 12,000 BC has ever been discovered in North America, through which the South American settlers must surely have passed. Recently, however, much firmer evidence of the existence of early settlement in South America has been found; charcoal from a hearth uncovered at Pedra Furada, in eastern Brazil, has been shown to be around 40,000 years old.

The advocates of this alternative view are of the opinion that the first people to arrive in America did not in fact do so during Beringia's final appearance above the waves, from around 25,000 to 13,000 BC, but during its previous appearance between 50,000 to 40,000 BC. Having taken one look at the imposing mountain ranges of the west coast and decided not to try heading inland, the settlers would naturally have stayed close to the coastline, having no real opportunity of spreading out into the interior until they reached the lower areas of the Central American isthmus. The same rise in sea level that drowned Beringia in 40,000 BC would then have swallowed up the coastal strip of North America down which they had travelled, together with all the traces of their occupation.

Whichever of these two theories is correct, it is clear that no single race of *Homo sapiens sapiens* was responsible for the colonization of the Americas. By 10,000 BC several distinct racial groups were already evolving. Although dental evidence collected from skeletal remains substantiates the view that native Americans all originated in northeast Asia, in the Siberian triangle formed by Beringia, northern China, and Lake Baikal, it also corroborates the theory that the original colonization consisted of three different waves. First there was a large initial migration, producing all of the native South American peoples and most later North Americans. Then, several thousand years later, came a second, smaller wave, that was made up of the ancestors of the Apaches and the Navaho, as well as the Indians of the Alaskan interior and parts of British Columbia. Finally, a third group of coast-dwelling peoples – the earliest Inuit and Aleuts – arrived to add their genes to the fast-expanding American pool. The basis for this three-wave theory is the striking similarity of the languages spoken by the different peoples in each of these groups; the linguistic evidence suggests that although, for example, the North American Indians are separated by an entire continent from the South American Indians, their

BOLD IMAGES OF WOMAN

Exaggerated in their proportions, yet conforming accurately to the essential features of the female anatomy, examples of statuettes found scattered through much of Eurasia testify to the feelings of awe and mystery experienced by their makers, who lived between 35,000 and 8000 BC. Most of the figurines are made of stone, baked clay or ivory, and measure between 3.5 and around 20 centimetres high. Known as Venus figurines, they represent women at every stage of life from puberty to old age, although the majority – including the ones illustrated here – are of mature and well-nourished females whose bodies show the effects of age and child-bearing. Archaeologists have variously speculated that the figures may have represented deities in fertility cults, or symbols of authority in matriarchal societies. But although they clearly had a vital role in the religious or social life of the hunter-gatherer communities who created them, their exact function remains obscure.

The protruding stomach and buttocks of this Italian stone figure are visible only when viewed from the side.

Angular lines distinguish this baked-earth Czechoslovak statuette.

Slender arms rest on the ample breasts of a limestone figurine found at Willendorf in Austria.

languages are close enough to indicate that they share a common ancestry and therefore a common date of arrival in America.

But these groups themselves began to diverge once they had settled down and embarked upon different life styles. From the pine forests of the Pacific northwest to the swamps of Florida, from the Great Plains to the Amazon jungle, from the woods of New England to the mountains of Peru, a wide range of habitats demanded widely differing survival skills. Isolated from each other, the inhabitants of such far-flung locations began to develop very different cultures, and, moreover, to develop a sense of what distinguished their particular group from people of "foreign" cultures. The idea of ethnicity was taking root. It had begun gathering momentum since the very early days of *Homo sapiens sapiens,* in Africa and Eurasia, but as time went on it would grow stronger and stronger.

It was not merely a matter of cultural difference, either: important physical variations were already marking the different races. The massive population increase, as humankind thrived across the world, was in itself a cause of change, each extra breeding person increasing the opportunity for genetic mutations and new ideas to occur. Any small group splitting off from a larger one would inevitably miss some components of the original group's genetic pool; over several generations, breeding within the new group strengthened and exaggerated these differences, with the result that the new group grew ever more distant from its ancestors. As *Homo sapiens sapiens* gradually spread out across the world, coming to grips with the wide range of different environments that he encountered, this process was reinforced by a process of natural selection, which ensured that those genetic variations suitable to a group's circumstances were retained, while those that did not were discarded. A tall and slender physique fitted the African for a life in the sunbaked bush, since it presented a large area of skin to be cooled by the air. However, the same body shape would have led to a prodigal squandering of heat in an arctic climate; the Eskimo therefore developed a much shorter and more compact form, and grew thick, straight hair, which helped to retain the heat.

Less than a hundred thousand years after making his first appearance in southern Africa, then, *Homo sapiens sapiens* had succeeded in conquering almost the entire globe. His was, in many ways, a tyrannical rule. From their beginnings as defenceless victims, human beings had evolved into wanton agents of destruction. Where once they had scavenged carcasses, or at best managed to kill the odd beast, they now drove vast herds to their deaths, with only a minimal amount of exertion. However, such a reign of terror could not last for ever, and, as many species were hunted to extinction and others became far rarer, major changes in strategy were required if there was to be enough food to sustain the growing population. Even during the ice age, there had been signs that humans were beginning to husband their resources. The hunter-gatherers who followed reindeer and ibex herds in Europe were beginning to exercise at least some loose control over them, corralling selected beasts and domesticating some for their hauling power.

Thus, when the ice age gave way to a new and milder era, around 8000 BC, the first stirrings of an agricultural revolution were already starting to make themselves felt at several different points around the globe. After two million years, throughout which they had instinctively followed the hunter-gatherer life style of their forebears, people all over the world had reached the stage in their evolution where they would start abandoning the spear for the ploughshare.

Nowhere is the hunter-gatherer life style more vividly recorded than in the many cave paintings and engravings of France and Spain. Created between 35,000 and 8000 BC, as the last ice age peaked and then subsided, this rich legacy of images provides a graphic testimony to the sensibilities of early human beings and to the extent of their cultural development.

Centuries of darkness have helped to preserve the freshness and vigour of these works. But brief glimpses of how *Homo sapiens sapiens* viewed the world about him raise at least as many questions as they answer. Were the images on the wall thought to contain a mystical or magical power? Are they symbols of some early religion or supernatural belief? Or were they simply used as educational aids – a kind of pictorial encyclopedia?

Three main subjects are portrayed, the least common of them being humans, who sometimes appear as stick drawings and at others as fanciful combinations of bird or animal and man. The clearest human images are handprints; the example shown on the previous page, which comes from the Pech Merle cave in the Lot district of southwestern France, perhaps served as an artist's signature or as an indication of territorial ownership.

Abstract symbols show up frequently, too. These take the form of lattices, boxes, barbs and rows of dots that may be an early form of hieroglyphics or an attempt to portray inanimate objects such as gates, huts, weapons or geographical landmarks.

The most frequently depicted images are of animals. These are mainly confined to four species – three of which appear in the painting below, from the Hall of Bulls at the Lascaux cave in France's Dordogne, with the addition *(extreme left)* of a mysterious unicorn-like creature. Here, as in most of the cave paintings, the animals are shown in profile only.

No-one knows for certain what the purpose of these images was. However, one clue to the mystery might lie in the inaccessibility of the paintings, which are often sited in alcoves and recesses only reached via a tortuous network of underground channels. Frequently, the caves that contained paintings were uninhabited, which suggests that the intention of the artists was not simply to provide a public art gallery. Also, many of the works are located in positions within the cave that would have presented great difficulty and discomfort to the artist. Far from being located at eye level on a handy, upright wall, many paintings are situated several metres off the ground, and would have required their creators to have daubed while either lying on their backs, perhaps on a rudimentary form of scaffolding, or else while clinging precariously to a rock face. At Lascaux, the artist or artists used the entire chamber for their composition, and incorporated the natural curves of the walls and ceiling to give more life to their images. The physical

obstacles involved in adorning this unusual canvas suggest a motivation running deeper than the simple desire for artistic self-expression or display.

One possible interpretation is that the paintings functioned as a sort of sympathetic magic. Most of the animals portrayed appear to be healthy, well-fed specimens, and a few even show signs of being pregnant, suggesting that the act of painting may have been intended as a charm, designed to promote health and fertility in the animal population and thereby also among the humans who fed off it. In the same way, by depicting beasts which are being killed with spears and arrows, the artists might have believed they were thus improving their own chances of enjoying success in hunting.

The role of the artist may have extended into the realms of the supernatural in another way. It is possible that painters were mediating between the hunters and the other world, conjuring up "spirit animals" on the wall in order to heal a wounded hunter or to help in the tracking of prey.

Such theories are reinforced by the apt suitability of these dark labyrinths to the staging of religious or ritualistic drama. Some experts believe that to early human beings the animal images bore a significance that was related not to species but to gender: one theory has it that the bison or aurochs represented femaleness and the horse maleness, their relative positions in a cave spelling out symbolic messages, perhaps about birth and reproduction.

Less ambitious is the idea that the artists used caves in the same way as modern-day teachers use a blackboard. Thus the contours of cave walls and ceilings might be employed by the artist to portray the surrounding landscape; also, apparently enigmatic rows of dots may have been intended to indicate the boundaries of a particular hunting territory. Paintings may simply have served as a means of informing fledgling hunters about how to identify various types of prey, in some instances, for example, specific features of an individual animal have been highlighted by a little twisting of the perspective, so as to display more clearly the detailed formation of an aurochs' hoof or the antlers of a deer. To spellbound children, crouching in a confined space and peering through the flickering flames of a hearth fire or a stone lamp fuelled by sputtering animal fat, such pictures would have stayed in the mind for many years.

The only sure thing seems to be that, although indecipherable to us, the messages conveyed by cave paintings were understood both by the people who painted them and by those for whom they were painted. Using the animal figures that appear on various cave walls, the following pages illustrate some of the evidence on which the different theories concerning cave art are based. But these constitute merely the pieces of the puzzle. How they fit together remains a mystery.

Two graceful galloping horses, both from caves in southern France, bear a distinctive M-shape on their sides, a device used to portray the change of colour between the dark hide and lighter belly area. Horses were the most commonly depicted animals in cave art, and these accurately executed paintings demonstrate the artists' familiarity with their subject. The creator of the "Chinese" horse on the facing page, so named because of its resemblance to horses in Oriental art, has given the animal a sense of movement and three-dimensionality by leaving the legs on the far side slightly detached from the body. The "circus" horse on this page, from Le Portel cave, takes its name from the lively high-stepping of its forelegs.

Peppered with dots, two horses stand back to back in a painting from the Pech Merle cave. Drawing on the ingenuity that these awkward work surfaces required, the artist has used the natural shape of the rock face to create a sculpture of the right-hand horse's head, while also painting the same horse's head in much smaller, more stylized form, as mirrored by the head of the left-hand horse. The dots and handprints were probably made by the artists' filling their mouths with liquid paint and then spraying or spitting it on to the walls.

A bold charcoal sketch from the Niaux cave in the French Pyrenees, shows a stocky, long-haired horse. Simple outline paintings dominated cave art, texture being added, as here, by the rough infill that gives the impression of a shaggy winter coat.

With tail lashing and hooves kicking, an enraged aurochs rampages past a pair of trotting ponies, at bottom right. This lively scene is taken from a Lascaux mural of considerable size: the aurochs alone measures 1.7 metres. The enigmatic lattice symbol at the left might signify a gate towards which the beast is being stampeded. The artist has created the impression of having used a wide range of colours, perhaps by painting in black on top of an older red image.

A bull charges across the rock face of a subterranean corridor at Lascaux. The difficulties of painting such a large portrait whilst maintaining accurate proportions were increased by choosing a site high on the cave wall. Traces of sockets dug into the walls two metres above ground level suggest that some kind of scaffolding, probably of branches, supported a platform on which the painters could work. The artist here has twisted the perspective of the front hoof to show its underside, possibly to teach hunters how to recognize aurochs prints.

Natural indentations made by water droplets form an eye and a series of flank wounds in this clever clay floor engraving from Niaux of a wounded bison. Most floor images have been worn away by human feet; this one was protected by an overhanging rock.

Two sculpted clay bison, each about 60 centimetres long, lean against some rocks in a remote, low-ceilinged chamber of a Pyrenean cave. The animals were carved on the cave floor out of blocks of wet clay before being propped against the rock face for the fine detail to be added. Once completed, the sculptures were never visited again, suggesting that the act of creating them formed part of a private ritual.

Four deer, sporting vastly exaggerated antlers, career across a wall of the Lascaux cave. They are exceptional for being outlined in red and yellow as well as the more usual black. Cave artists manufactured their paints by grinding natural earth and rock pigment to a fine powder, then mixing it with water to produce a paste or liquid.

A head to head confrontation between two ibexes is the subject of this bas-relief limestone carving, found in a cliff-face recess in France's Charente. Most ice age bas-relief work is found in the open air, but the reason for this remains obscure; the high standard of cave painting indicates that perfectly adequate illumination existed inside the caves.

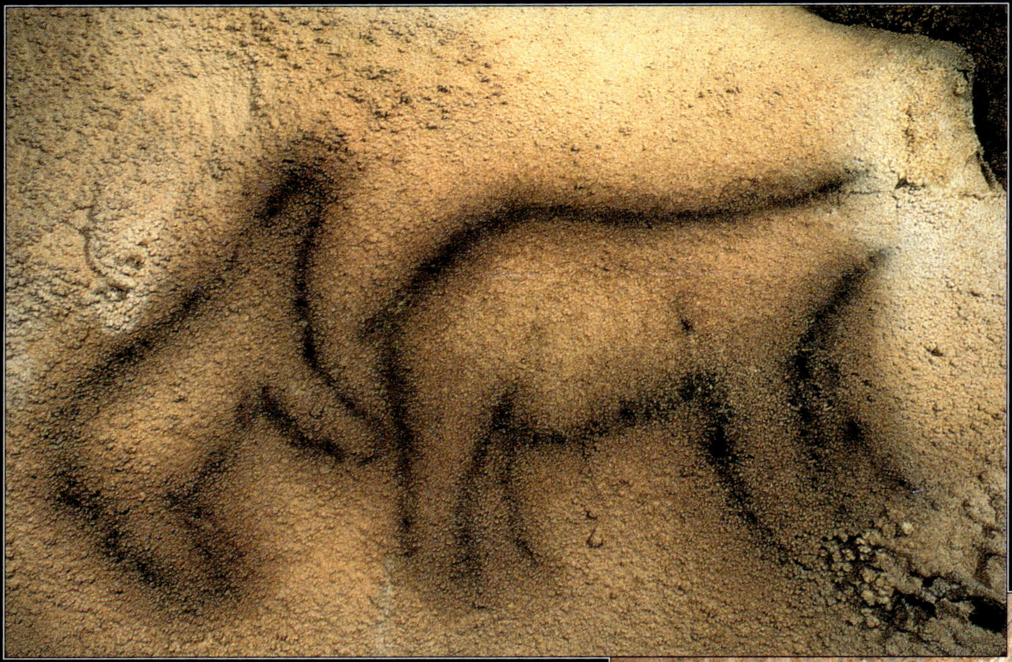

The image of a hair-coated mammoth on a wall at Pech Merle provides a reminder of the freezing temperatures in southern Europe during the last ice age. The abbreviated lines of the drawing indicate that this work was something of a lightning sketch, probably taking less than five minutes to complete. It forms part of a large mural featuring 25 animals and measuring seven by 2.5 metres.

Unusual for being depicted full face, the painted lion *(right)* peers from a wall at Les Trois Frères cave in southern France. Ice age artists rarely portrayed carnivorous beasts such as the lion, or the two headless bears *(above)* from the Ekain cave in northeast Spain, perhaps because close observation was too dangerous. Nocturnal creatures such as the owls *(below)* from Les Trois Frères also feature infrequently, again perhaps due to the lack of opportunity to study them.

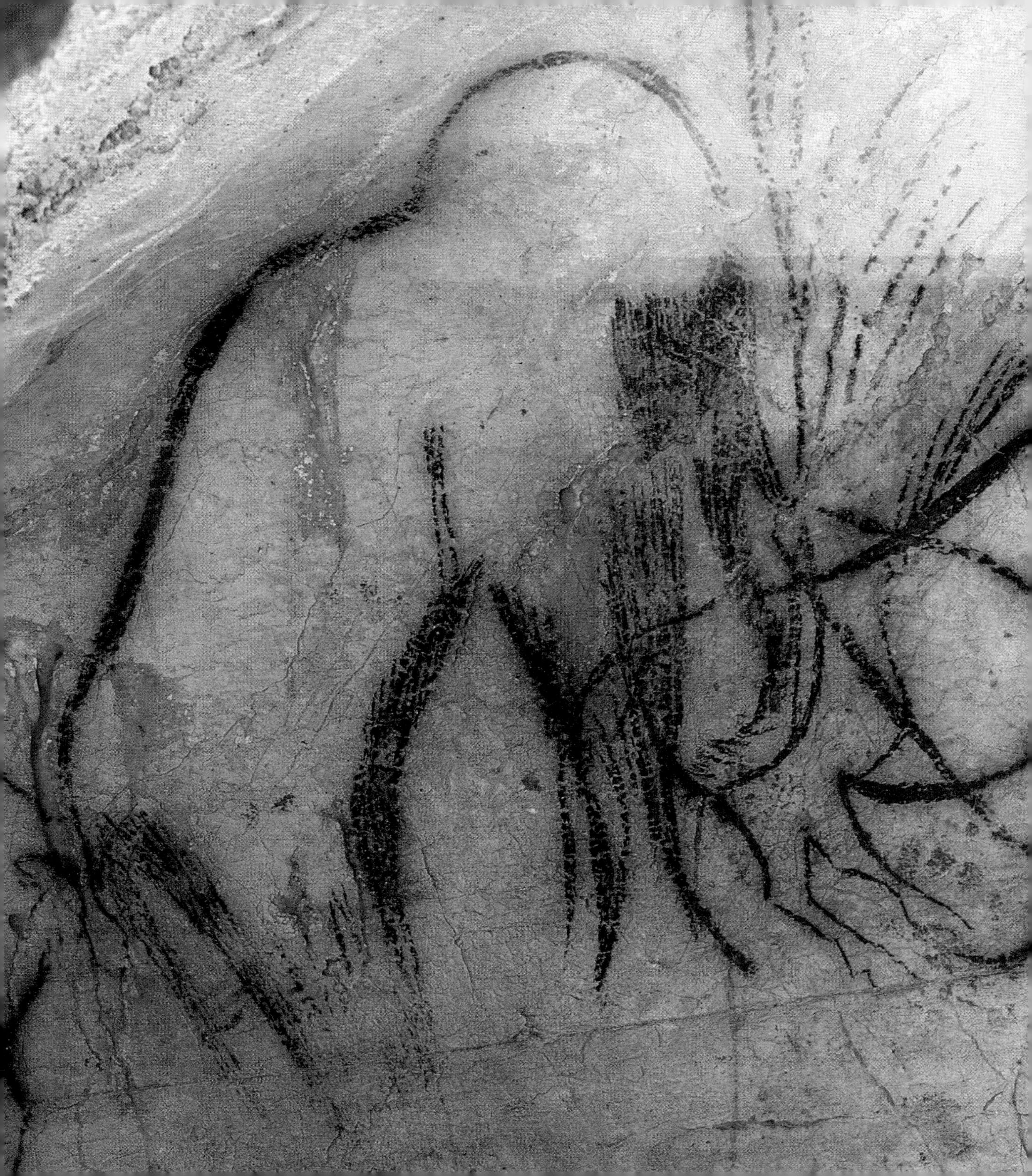

TAMING THE WILDERNESS

3

The camp at Beidha, in the Valley of the Ravens, had long been a favourite stopover for nomads wandering through the sandstone mountains of southern Jordan. It was a well-chosen site: set on a wide terrace, overlooking the wadi bed to the west, it afforded a good view of possible prey or enemies; water was available in nearby rock pools, supplied by a permanent spring which gushed from the crags 400 metres above the camp; and the low ground offered a plentiful supply of plant foods and wild animals. Year after year, small bands of hunters would halt there, clearing away the undergrowth and firming up the soft, sandy ground with layers of clay before setting up temporary shelters. It was not worth constructing a more permanent base, for when the surrounding countryside had been scoured of all sustenance — a matter of perhaps a month — the hunters would strike camp and move off to new pastures. After one or two seasons the game would return, the berries and plants would reappear, and the next group of visitors would arrive.

For centuries, Beidha had witnessed this transient passage of humanity. Then, shortly before 7000 BC, the rhythm of the valley began to change. For some reason the nomads who traipsed up to the terrace were staying there longer. On the site of the earlier huts sprang up clusters of round houses, whose floors were sunk half a metre into the ground. Above each pit rose a ring of sturdy, wooden posts, enclosed by plastered stone walls and roofed over with a clay-topped reed thatch. This was no short-stay accommodation. The buildings' very structures were designed for year-round occupation: the clay roofs kept off seasonal rains, while the sunken floors offered extra insulation during the cold winter months.

The new arrivals were too canny to think of themselves as permanent residents; there was always the possibility that drought, earthquakes or other natural disasters would force them to move on again. Besides, they would have been uncomfortably aware of one major drawback to their otherwise sturdy dwellings: the timber framework around which each house was built burnt well and frequently, often resulting in wholesale devastation of the nascent village. But from just such a tragedy would emerge a clue as to the reason for this sudden change in the valley's life.

On one occasion, as the posts charred and snapped, as the roof crashed down and the walls tumbled in, a family of fleeing householders discarded two precious baskets of food plants. Preserved beneath the debris these baskets contained a small scrap of carbonized knowledge that, 9,000 years later, would reveal Beidha's secret. When the site was excavated and subjected to the spotlight of modern scientific scrutiny, it was discovered that the baskets had contained grains of wheat. That in itself was nothing unusual: the plant grew — and still grows — as a wild grass in the region. But these grains varied in size and structure, indicating that important physiological changes were taking place. These changes were the result of the people of Beidha

A pottery figure of around 4500 BC clasps a broad-bladed farmer's sickle over its shoulder. Discovered at Szegvar-Tüzköves in eastern Hungary, this 25-centimetre-high statuette clearly portrays someone of bearing and importance; the regal, seated posture possibly indicates a deity. Agriculture first came to Hungary during the sixth millennium BC, having spread through the Balkans from its origins in the Near East. By the beginning of the fifth millennium BC, farming had been established in Europe as far north as the modern-day Netherlands.

cultivating their own crops. Planted in the valley's fertile soil and watered by the nearby spring, the grain would have provided the Beidhans with a ready source of food: they would have had no need to desert the valley in search of fresh forage. It would have been worth their while to build permanent dwellings — and to rebuild them after each conflagration. Agriculture had arrived in the Valley of the Ravens.

For two million years, human beings had been on the move, living off what they could forage from the wild. Villages such as Beidha heralded the end of all that. For throughout the Middle East similar settlements were springing up — and no doubt burning down — as other groups of hunter-gatherers strove to master the mechanics of farming: the sowing and harvesting of crops and the herding of domesticated livestock. These seemingly simple innovations were to change the face of the world. Within a few millenniums, the disparate mass of humanity had begun to coalesce into sedentary clusters that dotted every veld, steppe, plain and pampas on the globe.

It was not a sudden revolution. The habits of generations died hard and initially the old methods of hunting and gathering continued alongside the new way of life. Nevertheless, the pattern of human existence was irrevocably changing. For by bending nature to their will, the earliest farmers had done more than just increase their food supply. They had given themselves at least a chance — barring crop failure and disease — of living in greater security and to a greater age. They now had the option to diversify, to make

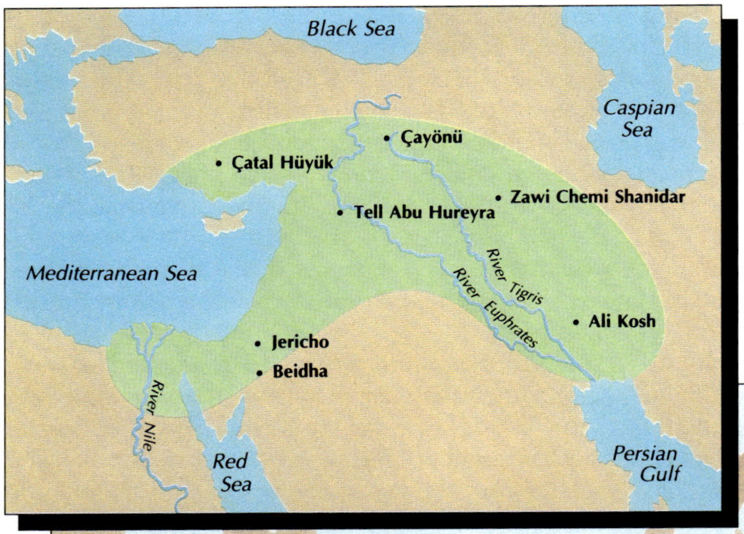

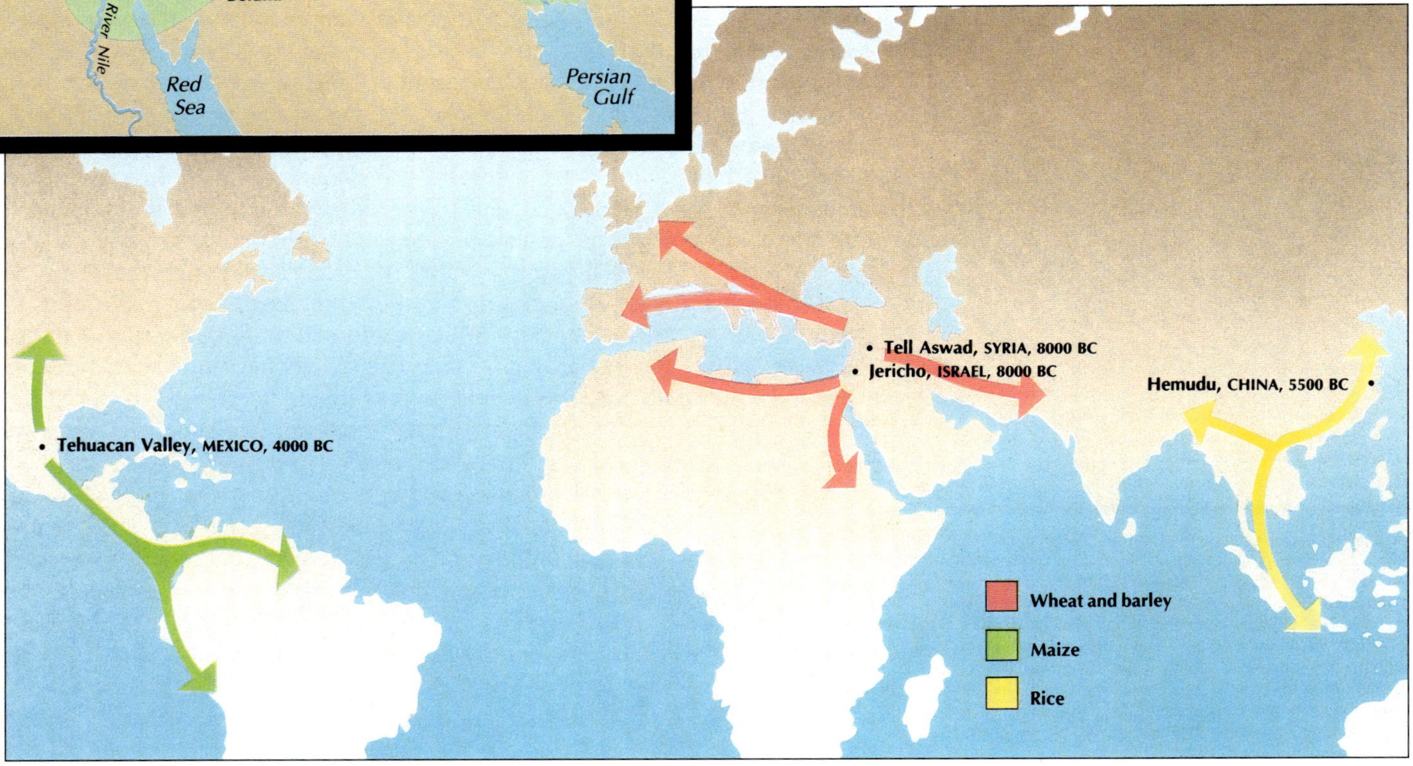

The main map shows the locations of the earliest known domestic crops, and the directions in which initial cultivation of those crops are thought to have spread. Much early crop domestication took place in the area of land known today as the Fertile Crescent *(inset map, shaded green)*. This region owed its fecundity to a warm climate and to the annual winter rains. The resulting abundance of plants and wildlife enabled hunter-gatherers to abandon their nomadic life style and to establish settled farming communities.

and trade desirable possessions, to accumulate wealth and to improve – indeed, to create – their lot. In short, they had gained some measure of control over their destinies, albeit in the process losing some of the social and economic flexibility that had accompanied their old nomadic life style. Once they had had a taste of this stability, few could resist it, and the settled, agricultural population steadily grew. Small villages became towns, towns expanded into cities, and these, in turn, became administrative centres for territories that would ultimately become empires. The foundations had been laid for civilization as we know it today.

Ten thousand years ago, as the frozen plateaus of the ice age shrank back to the North and South poles, the world came once more to life. Released from the climatic pall, vegetation spread, animals abounded and mankind multiplied. By 8000 BC there were between five and 10 million people on the globe, most of whom lived by traditional, well-developed skills in hunting and gathering. They scouted in small groups across the countryside, settling where the food was sufficient to supply their daily needs and moving on when sustenance became scarce. They were well aware of the natural cycle of reproduction, growth and death and the changing seasons. They knew their landscape well – where best to find grain seeds, roots and fruit, and where to hunt game at various times of the year. Material wealth was limited to what an individual could carry, but, provided the group was small enough, life could still be good: the meat they captured augmented their main staple of edible vegetation to provide a balanced and nutritious diet, they could grind grain into flour to make bread, and they knew which plants could be used to heal wounds and cure illnesses.

The thawing of the icecaps, however, had disrupted the established order. For, in spite of the hunter-gatherers' skills and intimate knowledge, their way of life best suited relatively small bands of about 25 to 30 people. As a group grew, it became more difficult to manage and to feed. When numbers became too large, family units were forced to split off to find new hunting territory – usually within the same region. As long as the population remained relatively low and land was freely available, this presented no problem. But with the earth's new fecundity, humanity was on the increase. In addition, huge areas that before had been dry land – such as large parts of the Mediterranean and Black seas, and later the North Sea – were submerged by the rising sea level, and their inhabitants were forced inland to find new areas to exploit. As a growing number of people competed for the richest and most accessible ground, it became increasingly necessary to consider a different form of subsistence strategy.

The first to do so were the hunter-gatherers of the Middle East. There, moist winds filtered eastwards from the Mediterranean to reach the Fertile Crescent, a strip of hilly but productive land that arched north from Jordan to the foothills of Asia Minor before dropping down along the Tigris and Euphrates rivers in Mesopotamia to reach the Persian Gulf. It was good hunting ground: on the upper ridges wild goats and ibexes found a favourable foothold, while on the lower slopes down to the valley floors sheep and gazelles grazed alongside wild cattle and asses. High on the hillsides, open woodland flourished with oak, juniper, hawthorn, pistachio and wild pear trees. More importantly, however, in those areas which were favoured by the sun and seasonal rains, vast fields of wild cereal grasses covered the ground, in some places spreading over thousands of hectares.

It was an abundance too great to pass by. Small bands of nomads would stop to set up camp beneath an overhanging cliff, in a suitable cave, or under rough shelters made of branches covered with leaves and grass. And as it gradually became clear that there was ample food for at least semipermanent occupation, so these rough camps became more substantial collections of huts. The densely packed stands of wild cereal — barley and two different strains of wheat, emmer and einkorn — were good providers. The wheat was of high quality, containing 50 per cent more protein than that used for bread flour nowadays, and each hectare yielded some 800 kilograms of consumable grain per year. During the three-week ripening season a family, stripping the stalks with their callused hands or with flint-bladed reaping knives, could gather enough grain to meet their annual needs and have some to spare.

There was not much chance of this bounty's diminishing, for the wild cereals proliferated with tenacious efficiency. Each seed was covered by a sharp-pointed husk along whose sides ran a series of backward-sweeping hairs, terminating in a few long bristles. As the ears ripened in the late spring they shattered, throwing their aerodynamic burden to the wind. The seeds scattered like darts, penetrating the soil point first, where their hairs held them in position; any disturbance by wind or rain simply made them wriggle deeper into the ground. During the hot, dry summer the seeds lay dormant, until the winter rains came to flush them farther into the soil to germinate and grow the next season.

Nevertheless, the evolutionary characteristics that had produced such plenty also militated against those who wished to enjoy it. Harvesting this brittle and elusive crop was an infuriating task. The ears of wheat and barley ripened fast to escape the unwelcome attention of predators, so timing was of the essence: too early a trip to any particular cornfield resulted in an easily gathered but unripe yield; too late a sortie might see the harvester standing frustrated amongst a sea of seedless stalks. Even if the moment were perfect, the slightest touch could send the precious grain scattering to the ground, from where its retrieval was a back-breaking business. Still more irritatingly, the individual seeds of each ear ripened at different times, so that every successful haul was likely to contain equal quantities of good and bad. Among the grasses, however, there were some that belonged to a mutant strain whose seeds not only ripened at the same time, but also remained on the plant until the whole ear was ready to drop to the ground. Standing out like beacons of plenty in contrast to the

Four maps show where livestock was first domesticated, and pinpoint sites where the earliest evidence of this has been found. These animals provided human beings with meat, milk and, later, hauling power, but they belonged to wilder, usually larger breeds than their modern-day descendants.

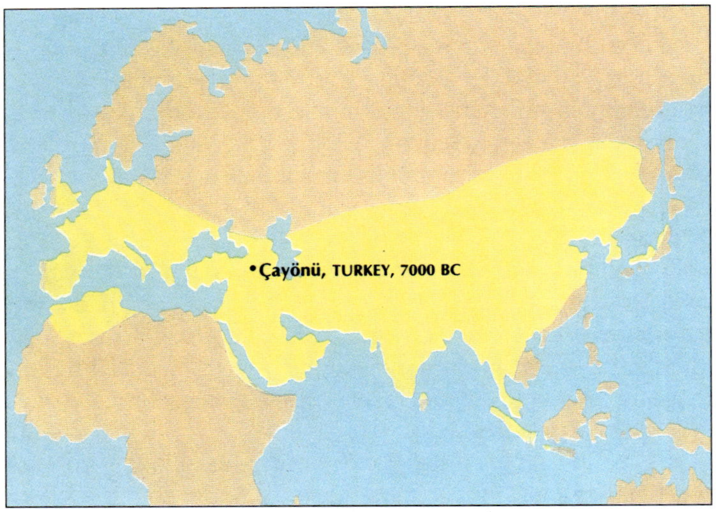

PIGS

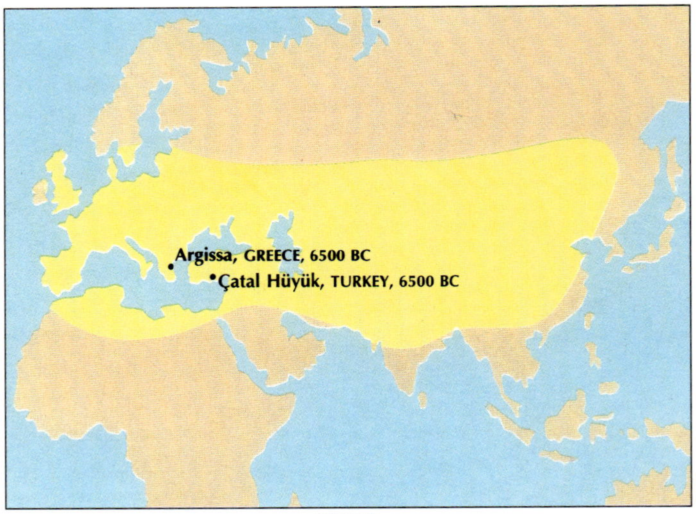

CATTLE

spindly outlines of their more dynamic cousins, these easily gathered and plump specimens would have been the first to be carried home in the harvesters' baskets.

Back at the camp, the day's yield would undergo the necessary processes to make it fit for consumption. First the seeds were threshed with sticks to remove the rough outer casings, then the resulting jumble would be tossed into the air to allow the light chaff to blow away, leaving only the nutritious kernels. Next, a sieving process would remove the smaller weed seeds and, finally, the end product was ground between stones in preparation for cooking. Amidst this flurry of activity some of the seeds would inevitably be spilt and swept on to the communal rubbish dump where they thrived untended. Eventually, these accidents of replanting would grow in number to the extent that they would form a new stand, not out in open countryside, but on the outskirts of the settlement itself, an arrangement that was both more convenient for gathering and easier to protect against scavengers. Almost unconsciously, the first steps had been taken towards crop cultivation.

It was a short step from opportunism to intent. Soon people began putting aside some of the precious non-shattering seeds, storing them in straw-lined pits in the ground or in pottery jars outside their homes, and then sowing them the following season to ensure another harvest of easily gathered grain. And gradually, over many years, farmers picked out and nurtured other mutant crops, thereby producing a number of changes in the make-up and character of the seed, amongst them an increase in both the size and number of grains on each ear. Despite the advantages of these new domesticated strains, wild cereals continued to be harvested. Their seeds' hard outer coating made it more difficult for birds to filch the grain in the fields and also gave the seeds a longer storage life. But the mutant grasses' greater productivity could not be ignored, and steadily the settled population began to expand.

Not only were the economics of the situation becoming increasingly obvious – about 650 square kilometres of land were needed to sustain 25 hunter-gatherers, whereas a mere 15 square kilometres could feed a community of 150 farmers – but village life also offered other advantages. There were fewer hunting accidents, for instance, and greater protection against carnivorous predators such as lions and leopards. Whereas small children had previously been a liability on the march, it

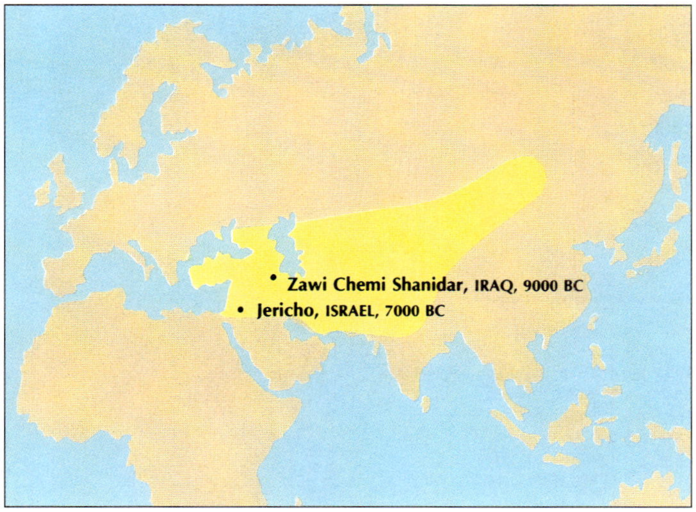

SHEEP

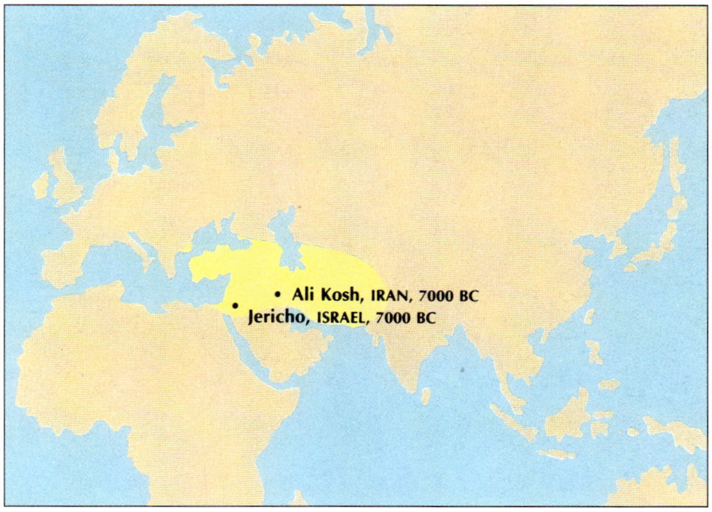

GOATS

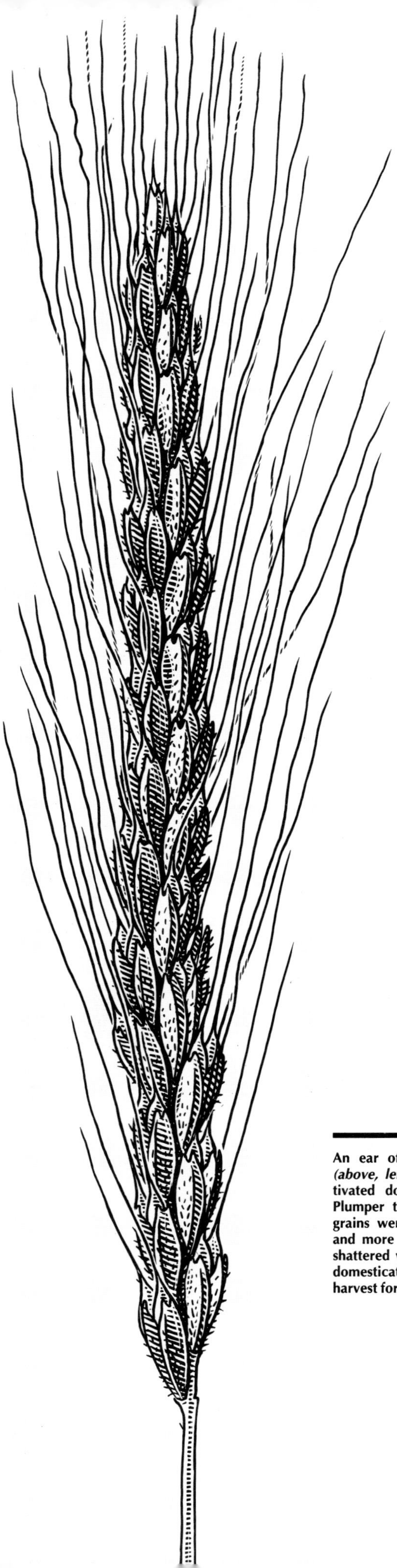

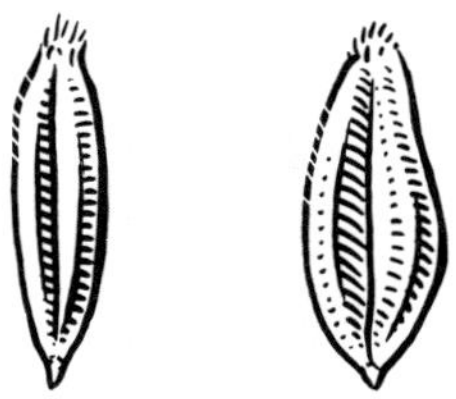

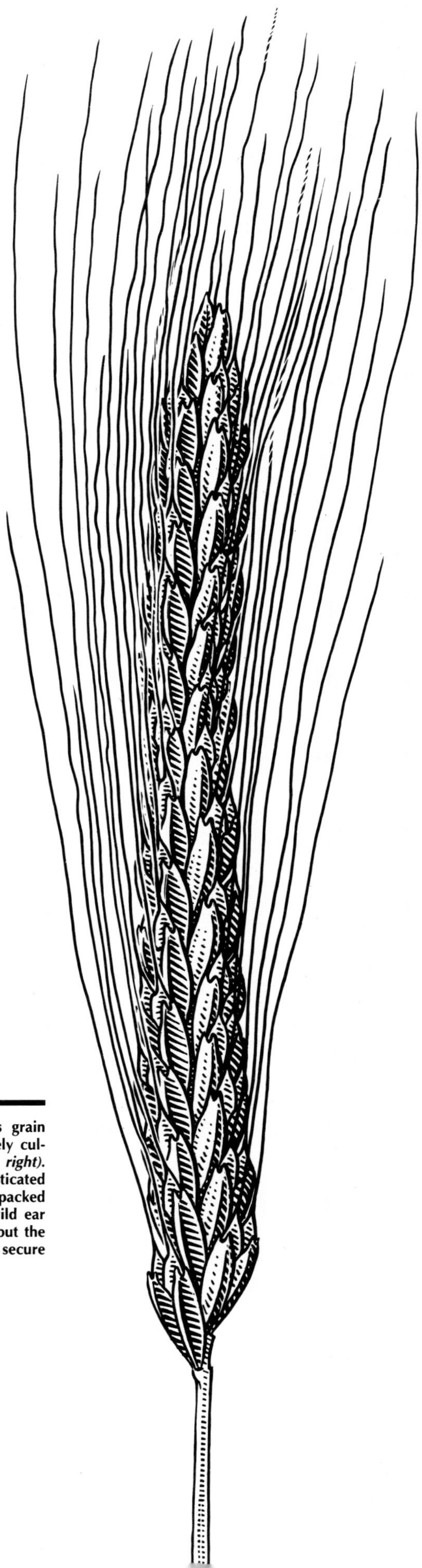

An ear of wild einkorn wheat *(left)* and its grain *(above, left)* differ markedly from the purposely cultivated domestic versions *(right and above right)*. Plumper than their wild counterparts, domesticated grains were encased by husks more densely packed and more firmly attached to the plant. The wild ear shattered when touched, scattering its grains, but the domesticated ear remained intact, yielding a secure harvest for early Middle Eastern farmers.

was now possible to raise larger families without fear of hampering the rest of the group. And the elderly, who before had often been an encumbrance, could now take a place in society, maintaining a continuity of experience, skills and wisdom. It was the older members of the community, too, who would look after young children, thus freeing the parents to work in the fields or to go hunting.

As the sufficiency of food turned into a surplus, so the settlers had both the spare time and the inclination to manufacture the kind of desirable possessions that during their nomadic days had been too much trouble to carry around. As a raw material – especially in areas that lacked suitable flint for making tools – they particularly favoured obsidian, a black, shiny, volcanic glass that was crucially important to both hunters and farmers alike. From a suitable chunk an artisan would skilfully chip off razor-sharp flakes for arrowheads, sometimes embellished with a barb and bound to the shaft with fibre, or glued in position with gum from an aromatic mastic tree. Smaller chippings could be set into a curved piece of wood to form a sickle, while large flakes would be shaped into knives and scrapers for cleaning meat from hides.

Nowhere, perhaps, were the advantages of settled life enjoyed more than at the large centre of Jericho. Situated beside a perennial spring that pumped over 4,500 litres of fresh water a day into the semidesert of the lower Jordan Valley, Jericho had witnessed one of the earliest instances of crop cultivation. As early as 8000 BC the surrounding fields were yielding a double and sometimes triple crop of wheat and barley which fed a population of perhaps as many as 3,000 souls. At the same time Jericho also supported the many peasants in the surrounding area, who came to barter obsidian, salt, semiprecious stones and luxuries such as cowrie shells and green malachite that could be fashioned into jewellery.

Long-distance traders had been operating for many years before the advent of farming. Since 9000 BC they had bought and sold obsidian along a well-established network which reached out from the volcanic areas of Acigol and Nemrut Dag in Asia Minor to encompass the whole of the Fertile Crescent. Grain, which was heavy though otherwise easy to transport in large quantities, may well have been one of the commodities in which they did business, and it is possible that these early merchants helped propagate the new agricultural way of life.

At the same time, however, the expanding farming population was unable to sustain itself. Not all settlements were as fortunately situated as that at Jericho. Most had limited resources, and as numbers grew and villages neared starvation point, so a small band – such as that which arrived at Beidha – would gather up a few baskets of precious grain and move off to new pastures.

The settlers at Beidha were just one small part of an economic migration that, by 7000 BC, had spread the process of selecting and cultivating grain throughout the Fertile Crescent. And it was not long before other plants were brought into cultivation. The wild, pod-bearing pea, for example, grew as a prolific weed, invading the grain fields and by the fourth millennium BC becoming a common source of protein. Over the years, from the large number of wild plants that had been filling humanity's cooking pots for thousands of years, a few individual species were gradually selected for cultivation – the ancestors of such familiar modern foods as carrots and cabbages, onions, olives, pears and plums.

In addition, the settlers soon made the important discovery that plants could yield more than just food. The seventh millennium BC saw farmers in the Fertile Crescent cultivating flax, not only for the linseed oil it produced, but also for its useful fibre.

Producing cloth from vegetable matter was a tiring process. First the flax stems had to be soaked, then vigorously pounded so as to release the fibres. Then the fibres were spun – either by rubbing them on the thigh or by twisting them on a free-hanging spindle – before being woven on a horizontal loom which was pegged out on the ground. The finished product, however, proved well worth all the trouble, being a fine linen that was both lighter and more comfortable to wear than the skins and woven animal hair previously used.

A further incentive to crop domestication was the discovery of fermentation. Probably some happy accident in the processing of barley grains revealed the secret of malting, and the potion it produced must have introduced a magical dimension to village life. Later, inhabitants of the Mesopotamian region of Sumer enjoyed at least eight barley beers, eight emmer beers and three mixed brews. The beverages were not always the purest: sometimes a straw was required to get beneath the layer of surface detritus. Nevertheless, the Sumerians rated the product highly enough to give it its own goddess, Nin-kasi, "the lady who fills the mouth".

Fruits, too, were cultivated for alcohol, but not before a great deal of trial and error and the discovery that a sweetener such as honey or nectar was needed to start fermentation. Regular cultivation of grapes, perhaps for wine making, was being carried on in the fifth millennium BC down the eastern flank of the Fertile Crescent. Vineyards in southern Mesopotamia were exporting wine to Egypt some time before 3000 BC, and at the time of Abraham, around 1800 BC, it is recorded that vineyards near Haran in northern Mesopotamia were producing wine from 29,000 vines.

Despite the joys of alcohol, the new way of life did not endear itself to all. Work on the land was hard, especially at the beginning of the growing year. First the farmers had to turn the ground with digging sticks – slender poles, about one metre in length, whose hafts were sometimes weighted with stone discs for added impetus. Then they had to burn the overcast from last season's harvest before the new crop could be sown; then, as the grain sprouted, unwanted weeds had to be wrenched out or removed by hoe and the field guarded against predators until the crop was ready to be gathered in. Only when food resources became scarce would people turn to this tiring existence, and for a long time crop cultivation merely supplemented the old methods of hunting and gathering.

The long-lasting importance of the hunting and gathering tradition can be seen at the village of Abu Hureyra, set on the banks of the River Euphrates in northern Syria. The Hureyrans were agricultural pioneers, having domesticated cereal crops at least as early as the inhabitants of Jericho. Nevertheless, for more than 2,000 years they continued to rely heavily on hunting. Meat was plentiful in the surrounding area: onagri – a species similar to the wild ass – goats, sheep, pigs and wild cattle roamed the steppes above the river valley, and every year the region was visited by gazelles which congregated on the banks of the Euphrates during the long summer drought. So rich a food source was the gazelle that by 8000 BC Abu Hureyra was home to a community of several hundred, or even several thousand, people, who derived 80 per cent of their meat requirements from this animal.

To slaughter a herd of these agile, light-footed creatures required a special strategy. Large-scale traps were constructed, usually near a valley through which gazelles were known to move. A trap was a wedge formed by two stone walls, set well apart at one end and gradually converging over a distance of several kilometres until they debouched into a circular, walled enclosure up to 150 metres wide. A migrating herd was stampeded and driven down this narrowing corridor into the enclosure. The entrance was then closed off and from concealed niches around the walls hunters would slaughter the trapped animals with a hail of arrows and spears. Alternatively, deep pits were dug behind specially lowered sections of wall. The gazelles leapt over the wall and fell into the pits, where they were quickly dispatched. The carcasses were processed at the killing site, the long bones being discarded and the meat salted and dried for storage and eating through the year.

So great was the migratory passage of game that huge numbers of these rudimentary abattoirs were constructed, running south for hundreds of kilometres across the killing fields of Syria. Some were linked in overlapping clusters to form chains that took whole herds at a time, and the great annual slaughter continued apace. But by about 6300 BC the devastation was beginning to upset the gazelles' instinctive pattern of migration and was even threatening the very survival of the species. As their main supply of food dwindled, the people of Abu Hureyra were forced to fall back on a different pattern of subsistence.

By then, however, man's knowledge of the animal world had been put to a less wanton use. As permanent settlements became more widespread, so their inhabitants sought to apply the benefits of domestication to the animals they hunted – most notably goats and sheep. These two species were ubiquitous all over the Middle East, thriving, respectively, in the mountain uplands and the rolling foothills, wherever there was vegetation and water. The more daring of a wild herd might even be seen within a village, snatching discarded food and browsing on rubbish dumps.

Two traits in particular made goats and sheep ideal candidates for domestication. Both were herd animals, used to staying in groups with a recognized leader, and both formed quick attachments to their parents at an early age. A young kid or lamb, separated from the herd – and probably suckled by a human wet nurse – would have soon bonded with its human parent. When it grew to maturity and bred, it would then

A primitive stone pestle and mortar, found in Iraq and dated between 7000 and 4000 BC, provides evidence that the first farmers of the Middle East developed techniques for turning wheat and barley grains into digestible forms of food. They used the grinding stone to break open the inner husks that contained the grain. Once freed, the grains could be winnowed to remove chaff, then mixed with water to make gruel or, after more grinding, used to make unleavened bread.

provide part of a tame and virtually self-supporting flock. In the wild these animals represented food, but in captivity they were important for providing wool for clothing and dung which was used for fuel and manure. Moreover, they increased farm output by being able to graze on marginal land unsuitable for arable cultivation. And when the mechanics of milking were mastered, it was not long before the early farmers' diet was being augmented by butter and cheese.

As they were with cereals, the settlers were quick to adjust evolution to meet their own ends. Within a few generations the species had been bred to become smaller and more docile. Even their horns became rounder and less dangerous – or, in some cases, disappeared entirely. Productivity, too, was boosted. Wild sheep, for example, which had come under human control as early as 9000 BC, originally possessed a stiff, hairy coat with a woolly undercoat that was shed annually. In time, domestication and crossbreeding produced an animal whose woollen fleece – in different shades and combinations of black, brown and white – was not shed each year but was sheared for the wellbeing of both man, who acquired a welcome extra layer of winter clothing, and beast, which got rid of an unwanted summer overcoat.

The beginning of the sixth millennium BC saw flocks of sheep and goats roaming at will through virtually every settlement in the Middle East. And by this time, too, the barnyard clamour had been swelled by the arrival of the pig. Fleshy, fast-breeding and capable of eating almost anything, the pig was an invaluable meat factory. When penned within an enclosed space it was prodigiously efficient at turning the soil, and such was its intelligence that when it was turned out into the surrounding countryside during the day, it would return of its own accord to its owner at night.

By 6000 BC, in the comparatively sophisticated centre of Çatal Hüyük, in southern Anatolia, the domestication of wild animals had broadened to include a less likely candidate: the aurochs, progenitor of modern cattle. Standing a good 1.8 metres at the shoulder, with long sharp horns, heavy muscles and an aggressive disposition, these were formidable beasts. That they came to be brought within the fold is a testament to human resolve and ingenuity. The male aurochs in particular would have been very difficult to restrain and it is unlikely that the smaller, but by no means docile, female would allow herself to be milked. They were difficult to manage and had to be kept from crop fields that could quickly be trampled and destroyed. But almost certainly there were religious as well as agricultural reasons for their capture. Amongst the many animals in the wild, the aurochs was a clear example of powerful strength and fertility. Its crescent-shaped horns were associated with the moon – which had a special place in the religious rituals of the people – and it may first have been captured for sacrificial purposes.

The braver farmers would have laid traps – probably leg nooses – for passing aurochs. Those who did not relish the prospect of trying to subdue a snared and no doubt furious beast would have chosen a less confrontational method. Their first step was to leave water and lumps of salt near a settlement. As the animals became familiar both with the site and the presence of humans they would have become more approachable. And from approachability it was but a short step to capture. The most amenable stock would be selected so that it could be encouraged to breed with the more docile cows to produce a higher milk yield, while those bulls not kept for breeding would be castrated to make them less ferocious. As a consequence of the process, selective breeding modified the characteristics of the species, reducing its size and in many cases changing the shape of its horns. Indeed, so great was the early stockbreeders' ardour to create a manageable beast that by the first millennium BC the shoulders of some breeds of cattle were no higher than one metre off the ground.

As well as rearing animals for food, man was also learning how to make them work for him. Almost certainly the early hunter-gatherers had struck up a semidomestic relationship with the wolf – to whom all present-day dogs owe their ancestry. A sharp-toothed and resourceful beast which competed directly with humans for available game, the wolf was a menace in the wild but an invaluable ally when trained. As animals that hunted in packs, shared their food and recognized the authority of a single leader, wolves were remarkably akin to humans, and when captured and reared from cubs they readily accepted life under their new masters. Initially used by hunters to flush out game, wolves soon became protectors and herders of livestock. Closer to home, they also acted as an efficient early warning against the approach of strangers and could even be used as pack animals to haul light loads.

It was longer, however, before other animals were pressed into human service. Not until at least 4000 BC or later were the horse, camel and ass tamed to provide a means of transport; and a millennium later, cats came off the refuse dumps to take on the role of household rat-catchers.

Once people had become settled in one place for long periods of time, they began to build more substantial houses to accommodate themselves and their growing families. By 6000 BC there were small townships in central Mesopotamia that had expanded to cover areas of some four to five hectares with houses of mud-brick walls roofed over with timber and plaster. Each of the dwellings would have been equipped with a hearth for cooking the daily meal – a cereal gruel or a meat stew, perhaps, supplemented by boiled roots, bread, fruit and nuts and possibly a thick porridge made of

Carved on a lapis lazuli cylinder seal *(left)*, a Sumerian drinker of the third millennium BC enjoys beer from a deep bowl. In order to reach through the thick scum that formed on the beer's surface during the brewing process, he uses a long straw, although it may not have been as precious an item as the gold straw pictured below, which measures 1.36 metres and belonged to a high-ranking priestess in the temple of the moon god Nanna, at Ur. A drawing showing more of the seal *(above)* reveals that beer drinking was a communal affair.

wheat grain. The hearth also provided an important source of warmth for the family as they huddled under their sheep or goatskins during the cold winter nights.

Beside some of the houses stood outbuildings used for long-term grain storage. To others were added small rooms where settlers could make weapons, tools and personal adornments for their own use or for barter. Beidha, for example, boasted a six-metre-long row of workshops, the building providing work space and upper-floor living quarters for six artisans. Craftsmen would specialize in different materials: a boneworker might make such objects as pins, beads, necklaces or shoulder-bone shovels; stoneworkers would produce agricultural tools, grinding stones and mortars and pestles. In Beidha there was even a butcher and a beadmaker.

The most ubiquitous craft, however, was that of the potter. Clay vessels had been constructed by Japanese hunters and gatherers as early as the 11th millennium BC, but their very nature – both fragile and heavy – made them impracticable. As a more permanent way of life was adopted in the Near East, pottery jars came into their own as essential equipment for storage and cooking. The techniques were simple, requiring neither special skills nor tools, and virtually every member of society could make his own pots. Containers of all shapes and sizes were fashioned by rolling long ropes of clay that were then curled around, one layer on another, to whatever height was required. Other vessels were made by simply kneading a ball of clay into shape with bare hands, or by forming it around a mould made of stone or wood. The vessel was then smoothed by hand before being baked at a temperature of around 800 to 900 degrees Celsius on an open fire.

By the fourth millennium BC, the introduction of the potter's wheel and high-temperature kilns had made pottery a specialized occupation, with just a few skilled craftsmen supplying the needs of a whole community. But even before then the craft had developed beyond the simply utilitarian and become a means whereby villagers could give their artistic expression full rein. Pigments were made from substances such as ochre and malachite to brighten the basic pottery vessels. And decorative elements were incorporated into the design, often inspired by familiar subjects such as animals, human beings engaged in various activities, and religious symbols to appease the gods or ward off evil spirits. Toys were also constructed for the amusement of children: bracelets, bangles, necklaces and models of people and animals.

Such modelling expertise was also pressed into the service of religion. There is little doubt that man's intimate association with nature led him to believe in and respect some external, invisible power that regulated the whole cycle of life: its growth, reproduction and death. Many thousands of small figures handworked in clay were produced, representing symbols of fertility – naked female figures, enormously fat and with exaggerated breasts and thighs, some in the position of childbirth.

In Mesopotamia, collective worship became institutionalized for the first time shortly after 5000 BC in the town of Eridu, southwest of Ur. In honour of their patron deity Enki, who presided over the town's water supply, the people constructed a substantial building that was both a temple and a focus for the town's agricultural and manufacturing activities. Over the years, as each building fell into disrepair, another was built upon its foundations, time and again, one upon another, in its later stages taking the form of a high, stepped tower or ziggurat. This architectural form was to be repeated many years later in other parts of the world.

The temple was the hub of the town's life, where surplus agricultural and hand-crafted goods were centred and distributed under the auspices of the incumbent god.

In the bustling marketplace, packed with sellers whose vegetables and handcrafted wares lay spread out in front of them, the gossip and news that passed from one to the other were as important and valuable as the items that were traded between them. Amidst the exchange of knowledge, however, conversation must occasionally have turned to darker matters: the price in death, destruction and disaster that the early farmers paid for their success.

For the new way of life that had brought so many benefits and solved so many problems had its disadvantages too. The stability of settled existence was constantly under threat from the vagaries of nature, such as epidemic disease, flash floods, violent dust storms, scorching winds and insect plagues. Moreover, it was not only natural hazards that beset the early farmers. The amassed wealth of a community was an obvious target for marauding groups of nomads or envious neighbours, and increasingly the spears, slings and arrows which previously had been used mostly against animals were now directed towards humans. Designed as it was to deal with such powerful creatures as the aurochs, the hunter's arsenal was devastatingly effective. A sling, for example, had a range of up to 180 metres, and could launch a stone the size of a fist, or a hardened clay ball, with enough force to smash bones. If they did not post a guard, the inhabitants of an unprotected village were easy pickings: all night-raiders need do was stand outside the doors of each house and slaughter the settlers one by one as they emerged. Accordingly, one of the first priorities for any new settlement was to build a protective enclosure.

Health was another problem facing the villagers. As settled farmers producing relatively few forms of food and with less time to forage for the wide variety of natural foods the human body requires, they inevitably suffered deficiency diseases. In addition, the accumulated filth of human settlement brought the bane – unknown to nomads – of food-contaminating vermin such as rats and cockroaches. Constant close contact with livestock produced a range of new ailments: measles, influenza and smallpox were just a few of the viruses transferred, in one form or another, between animals and humans. Within larger centres, a high density of people living in close proximity with limited sanitation would have been prey to the spread of many infectious diseases such as dysentery, typhoid and tuberculosis – a situation often aggravated by a settlement's geographical location. At Çatal Hüyük, for example, situated in a marshy lowland basin, malaria was rife and may have caused many early deaths. Very few farming people in 6000 BC lived to middle age – 30 years for women and 35 years for men was the general life span they could reasonably expect.

For all the drawbacks of settled life, populations expanded rapidly, and the need to grow more food soon forced farmers to increase the amount of land under cultivation. One way was by irrigation. It would hardly have escaped the attention of early settlers that the most productive tracts of land were those that were soaked each time the streams or rivers flooded. From there it would have been a short step to damming the flow and diverting the water through a ditch to reach drier land. This proved particularly fruitful in the plains of Mesopotamia, where the floods of the rivers Tigris and Euphrates laid down fertile deposits of alluvial silt in areas where the natural rainfall was insufficient to support crops.

It was easy enough in principle, albeit laborious in practice, to construct the irrigation system. Where the rivers flowed across flat plains, years of flooding and the continual depositing of silt had caused the water to run, within raised levees, above the level of the surrounding countryside. It required only a primitive shovel to cut through the river banks and form the ditches, plus a basket to carry away the earth. Each ditch that sprouted from the river was in turn connected to numerous minor channels, so that eventually the ground was laced with an intricate web of waterways. Although simple to build, such a network demanded continual attention. The ditches had to be cleared regularly of their accumulated silt, and boulders or wooden boards positioned to divert water through the various channels.

As well as bringing life to previously barren land, the villagers also learnt how to intensify cultivation on existing fields. The digging sticks that early settlers had inherited from their nomadic ancestors soon gave way to a more effective device, formed from a J-shaped piece of wood or sapling, and known today as a scratch plough. To use it, one farmer hauled on a length of rope attached to the longer end of the piece of wood, while another plunged the shorter section into the earth, directing it across the field to leave a shallow furrow.

It was longer, however, before the early farming communities recognized the full potential of the powerful cattle they had taken so much courage and effort to domesticate. They may have harnessed the strength of oxen to pull large boulders from their fields or to shift fallen trees, but it was not until the middle of the fifth millennium BC that agriculture experienced a significant change, brought about by the application of animal strength. With an amenable, trained beast harnessed to a plough, farmers could not only turn fields far quicker than before but could also open up ground too heavy or too weed-choked to be cleared by hand.

The first proper plough was a rudimentary affair: a single stout stick, pulled by rope attached to an ox, scratched shallow seed-rows in the soil. It may not have been until the third millennium BC that farmers in Egypt and Mesopotamia produced a more sophisticated design: a heavy piece of pointed wood to penetrate into the soil, with a crossbeam connected to the ox by ropes, two handles for steering the plough in a straight line, and a funnelled tube immediately above the ploughshare that allowed the ploughman to pour seed directly into the furrow, treading it in as he went along.

Other methods of increasing food production were spurred more by necessity than design. Year after year the early farmers planted their grain, and inevitably the soil became increasingly impoverished. They may not have understood why the annual crop was diminishing – that the nitrogen, phosphates and potassium that growing vegetation requires from the soil were becoming exhausted. Whatever the reasons may have been, they would often abandon the stricken field and clear a new space for cultivation. Livestock was turned out on to a piece of nearby open woodland to crop as much of the standing vegetation as possible, while at the same time nourishing the soil with their droppings. Trees were felled with stone axes, or killed by means of ringbarking – cutting around the trunk's circumference to stop the flow of sap – before being cut into manageable pieces and taken to the village for firewood. Reaping knives dealt with bushes and shrubs, while digging sticks were employed to prise out roots and boulders. The work was done in the hot, dry months of the year, before the rain came and the seed would have to be planted. The timber and debris drying under the hot sun were burnt, leaving ash and dying vegetation to enrich the soil. Cattle were again driven over the clearing to trample the ground as the next season's grain was scattered. The new crop would be carefully watched over as it grew – after the sweat and toil of so many people it could not be allowed to be demolished by straying animals or choked by unwanted vegetation. The new season's harvest, rich and plentiful, more than paid for their united efforts. Meanwhile the old, abandoned field was ignored as it slowly reverted to its natural state, steadily regaining its fertility through decaying plants and rain-borne salts until it was ready for use once more.

As long as there was enough fresh land to cultivate to keep pace with the increasing size of the community, this cycle of abandonment and expansion was effective. But in areas restricted by unsuitable growing terrain, enclosed by hills or with poor water drainage, the system had serious limitations. It took 20 to 25 years of lying fallow before an abandoned field could produce another healthy crop. In many settlements where land was scarce, the fallow fields had to be brought back into use too soon. To compensate for lower crop yields the villagers increased the quantity of their livestock, often above the limit their grazing could safely sustain. Overgrazing, continual use of the same fields, and possibly a few seasons of drought led to the start of desertification of once luxuriant regions. It was enough to force villagers to abandon their settlements and find a new location, as their forefathers had done centuries before.

Farming soon spread westwards to Europe. Here the waning of the ice age had brought problems as well as plenty to the local hunter-gatherers. Previously, much of the landmass had been a cool steppe that was home to herds of horses, deer and bison. The existence of the giant deer, which boasted an antler span of as much as 3.5 metres, was

VESSELS FOR THE HOUSEHOLD

The making of pottery containers came only with the development of settled communities; weighty and breakable objects were of no use to earlier nomads. The first known pots were made in Japan around 10,500 BC: coils or lumps of clay pressed together by hand and hardened over a bonfire or baked in a covered pit. By 3500 BC, potters in Mesopotamia were using stone or wooden turntables to rotate their pots during the moulding process – as depicted in the limestone statuette of a working potter *(right)*, found in Egypt and dating from between 2500 and 2200 BC – and purpose-built kilns to bake them in. Central pivots later enabled the turntables to spin more quickly, thus speeding up the production process. The examples shown on the following pages chart the progress of the early potter's art.

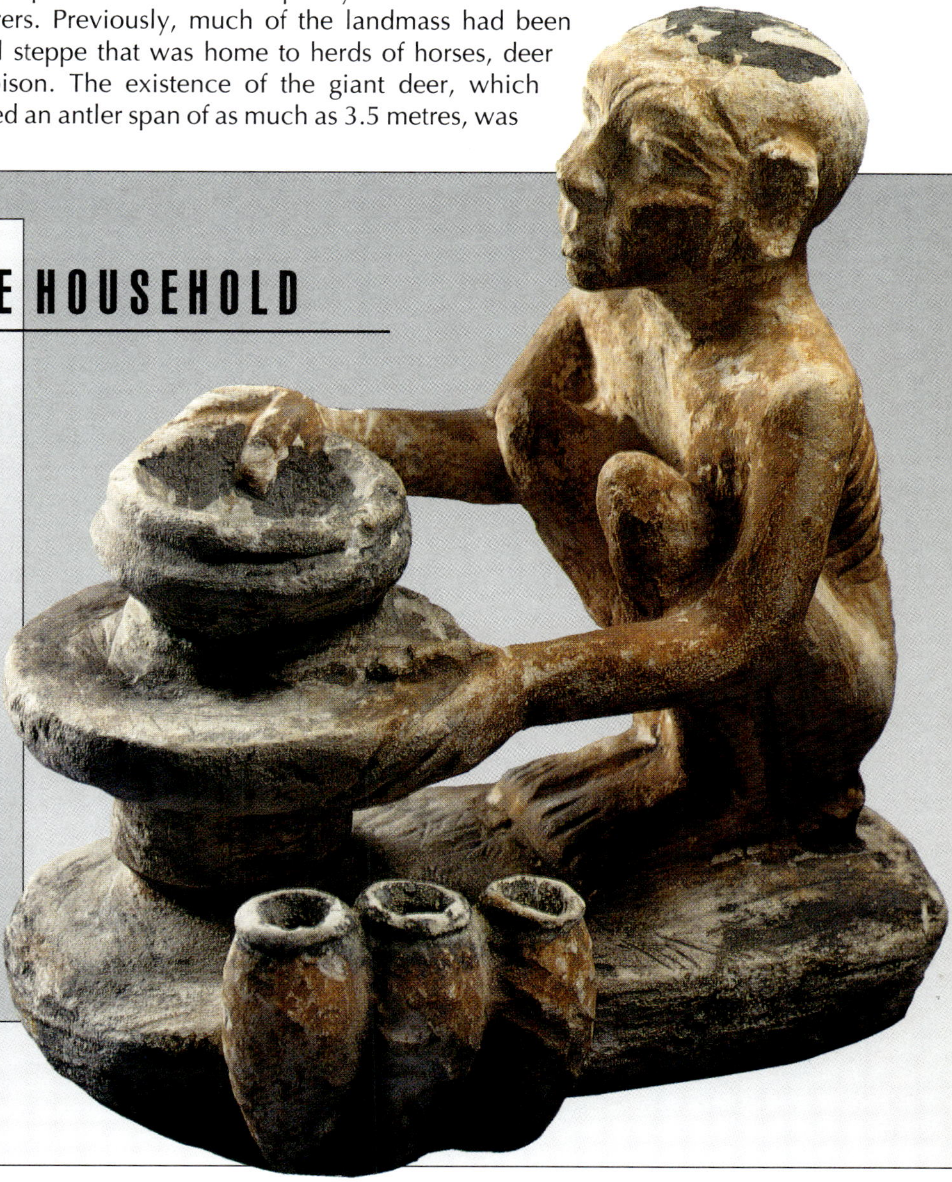

just one testimony to the lushness of the grazing. As the glaciers retreated, however, Europe had been colonized by woodland, first hardy specimens such as pine and birch and later the warmth-loving oak and beech. Amidst these forests, hunters could no longer find whole herds, but were forced to flit between the trees, picking off individual beasts with spears and arrows. To the north, a hardy few still clung to tradition, culling the migratory reindeer as they plied their annual round or skimming across the frozen wastes on skis and dog-drawn sleds after herds of elks. But generally the best living was now made by fishermen, who paddled their skin-covered coracles across the continent's rivers and coastal estuaries.

The new farming techniques brought from the Middle East wrought transformations on this timbered landscape, rapid in some areas, slow and gradual in others. By 6500 BC agriculture had been introduced to the Balkans; around 500 years later it was established on the Italian peninsula, and another millennium or so saw settlements springing up in eastern and southern France. It took some considerable time to develop new strains of domesticated grain that were able to thrive in the cooler and moister conditions of northern Europe, but by 4000 BC successful cultivation had found its way northwards through Hungary, reaching the Netherlands and Poland.

A pointed-base Japanese cooking pot from the sixth millennium BC may reflect the earlier use of animal horns as drinking vessels by the hunter-gatherers who made it. The pot bears the rope-pattern or *Jomon* marks that gave their name to the period in Japan from 10,000 to 300 BC.

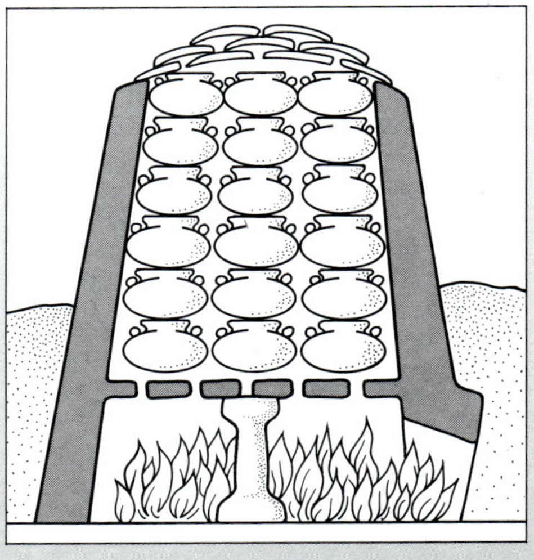

Clay pots are fired in an early Near Eastern ceramic kiln in this cross-section diagram. A ventilated clay floor separated the pots from the flames, making cracking less likely. On top of the kiln, loose shards gave insulation; a draught tunnel at the base allowed for temperature control.

At times the forests were a hindrance. They made travel slow and grazing difficult. Settlers would have to cut down entire tracts of woodland in order to make clearings in which they could pasture their sheep and goats. At other times the trees were a boon: not only did they prevent soil erosion, but they also provided a plentiful building material. In woodland clearings and along the river basins whose rich loess, or alluvial silt, provided fertile growing ground, Europe's farmers constructed wooden settlements. Poles were sunk into the ground to form the frame of thatched houses, as much as 40 metres in length, whose walls were composed of wattle and daub – thin, interwoven switches covered with clay. Each dwelling was divided into three sections, with living quarters sandwiched between a cattle stall and a lofty storage barn. Ditches and earthen banks protected the settlement from human predators, while wattle fences safeguarded crops from animals.

The new climate that had welcomed agriculture into Europe was a fickle host. The plentiful rainfall meant that, unlike in the Middle East, crops would grow almost everywhere without the need for any irrigation. On the other hand, too wet a season

The lack of decoration on this Mesopotamian pouring vessel, dated between 3700 and 2850 BC, suggests it was destined for everyday use. The introduction of pivoted turntables made it possible to mass-produce such complex shapes.

could flood large areas of river valley and low-lying coastland. In a bad year, for example, the River Danube might rise by as much as four metres, causing its tributary, the Tisza, to flow backwards, inundating farmland for 100 kilometres of its length. It would have been scant comfort to Europe's riparian dwellers, as they apprehensively watched water levels rise, to know that elsewhere the changing weather pattern was having an even greater effect.

In Africa the end of the ice age had had a dramatic impact. The Sahara Desert, previously too dry and inhospitable to sustain life, now experienced periods of heavy winter rain blowing in from the Atlantic. The desert shrank to a narrow strip, surrounded by wooded uplands and grassy plains in which human hunter-gatherers competed with new indigenous fauna such as cattle, giraffes and elephants. New lakes and rivers appeared, while earlier water levels changed radically. In 12,000 BC Lake Chad had not existed, but within 6,000 years it had swelled to cover over one million square kilometres – an area more than 10 times its present size. At one point the waters of Lake Turkana in Kenya swelled so dramatically that they overflowed into the Nile watershed, and the River Nile itself rose to such remarkable heights that one spectacular flood deposited 20 metres of silt on the Sudanese flood plain of Wadi Halfa.

It was the prodigious flooding of the Nile that was to form the basis of Egyptian agriculture. Set in a narrow, flat-bottomed valley, the northern stretches of the river were surrounded by arid territory. Every September, however, the river would burst its banks, flooding the valley floor before retreating the following month, leaving in its wake a layer of fertile silt. This moist earth was ideally receptive to domesticated wheat and barley, which reached the region from the Fertile Crescent in about 5000 BC. And as the population grew, the Egyptian farmers soon learnt to capture the annual flood water and divert it towards those areas that remained relatively dry. Low, clay-lined stone dykes were constructed to retain the retreating October waters, which were then diverted by sluices and canals throughout the whole valley. By 3000 BC the valley was a chequerboard of carefully administered irrigation basins whose ample produce succoured one of the most successful civilizations of the time – the Egypt of the Pharaohs.

Elsewhere in Africa, agriculture got off to a slower start. Shortly after the arrival of wheat and barley in Egypt, the crops were being grown by nomads in the east of the Sahara, probably along with sorghum, whose large flower heads ripened

Geometric designs adorn this fourth-millennium-BC goblet from Susa, in Iran. Trading in painted pottery began in the Near East around 6000 BC, centred on northern Mesopotamia and Iran, where artists used red and black pigment to create floral, human and animal motifs.

to glossy seeds that produced a syrup or could be ground into flour. Livestock began to be herded, and agriculture gradually established itself as far west as Senegal.

For nearly 2,000 years these pastoralists of the Sahara settled to a regular pattern of growing at least some of their food and herding cattle, yet still hunting and gathering. They were an artistic people, who used catfish spines to decorate their pottery with distinctive wavy lines and decorated exposed rock faces with paintings that vividly illustrated the animals so important to their lives. But by 2000 BC the first flush of postglacial warming had worn off and the Sahara reverted to the arid wasteland we know today. The lakes dried, the grasslands withered and the herdsmen retreated, leaving their art to posterity and the windblown sands of the desert.

Those who headed northwards would have found a fairly familar environment in the mountains along the southern coast of the Mediterranean. For those who emigrated south, however, tropical Africa offered a completely different scenario. Here the extremes of climate favoured a vegetation that was a far cry from that which grew in the Middle East. In modern-day Nigeria, for example, the groundnut and the pigeon pea were probably cultivated to sustain human settlement, as were the plentiful yam and the *Ensete*, or false banana, the stem and root of which provided fibre and medicine in addition to food.

The domesticated crops from the north soon could not gain even a tenuous foothold in the woodland clearings, river banks and forest margins where the native plants thrived. This time, the expected agricultural boom never occurred. Despite the apparent lushness of the region, the newcomers found that the environment was often unsuitable for crop cultivation. Wheat and barley, so important in western Asia, would not grow; they needed winter rain, and in the tropics the wet season was the summer. There remains only scant evidence of any indigenous crops being cultivated too, though this may be at least partly due to the fact that root plants, such as the yam, leave virtually no archaeological remains, whereas traces of seed plants, such as grasses and cereals, often survive either in the form of carbonized grain or as seed-shaped impressions left in pottery.

Poor soil also hindered the would-be farmers. Since the dawn of time, the warm rains that doused the tropics had been washing essential nutrients out of the ground. Even the flinty silica, a rocky residue that helps retain nutrients and lends a grey colour to soil, had been removed to leave the earth a barren red. Only in the tropical forests was a suitable reservoir of nutrients retained. But as farmers slashed and burnt their way through the trees they merely cleared the way for further impoverishment. In just a few decades the red soil would once again be washed out and infertile.

While being denied the carrot of good yields, the farmers of tropical Africa were also spared the stick of population pressure. Here disease, most notably sleeping sickness spread by the tsetse fly, was prevalent wherever a forest clearing was opened up for cultivation. And where irrigation had been so successful in places like Egypt and southern Mesopotamia, the introduction of standing water and irrigation ditches in tropical Africa resulted in a widespread incidence of river blindness, bilharzia – a disease carried by water-borne parasites that attacks the internal organs – and malaria that had a marked restraint upon the numbers of both humans and animals.

The savannas and forests of sub-Saharan Africa nevertheless offered huge areas into which farming could expand. With the pressure of disease keeping numbers down, in such a vast and empty land small family groups could split off from the main community and create their own villages. This steady migration to new locations

A hollowed-out hearth forms the centrepiece of this typical six-by-six-metre Skara Brae house. At the back stands a stone dresser, on which the family could display prized objects – perhaps a necklace made with the teeth of sea creatures, a stone ornament or the two bone objects below, possibly used as dice.

THE STONE HOUSES OF SKARA BRAE

The inhabitants of Skara Brae, a small, third-millennium-BC farming community on Pomona, the largest island in the Orkneys, off northern Scotland, lived in stone houses built inside mounds of solidified refuse. They hewed their homes from a great dome of midden, a mixture of vegetable matter, dung, shells and stones, which, when left to decay, eventually acquired the texture of toughened clay. Within this they fashioned their living quarters, joined by interconnecting passages and lined by dry-stone walls on both the inside and outside to provide a sturdy, weatherproof environment. No trees grew on the island, so the people of Skara Brae probably used driftwood, supported by whalebone and thatched with turf, to roof their homes. The scarcity of wood also meant that they made their furniture out of stone, using bracken for bedding and animal skins for blankets and curtains. Despite their somewhat spartan surroundings, these early Orcadians had a varied diet. On smoky fires, fuelled by a mixture of dung, whalebone and heather, they cooked a number of different types of meats – beef, lamb, pork, venison and goat – as well as a wide range of seafood and even sea birds, including the gannet, guillemot and great auk. Fresh water was supplied by nearby streams, and milk was provided by their herds of cows and goats, and flocks of sheep.

A ground plan outlines the shape of Skara Brae in 2800 to 2500 BC, itemizing the interior features of the house pictured opposite and demonstrating the marked similarity in size of all the houses. Only one structure, on the far right of the plan, stands apart from the rest; stone fragments littering its floor indicate that it may have served as a workshop.

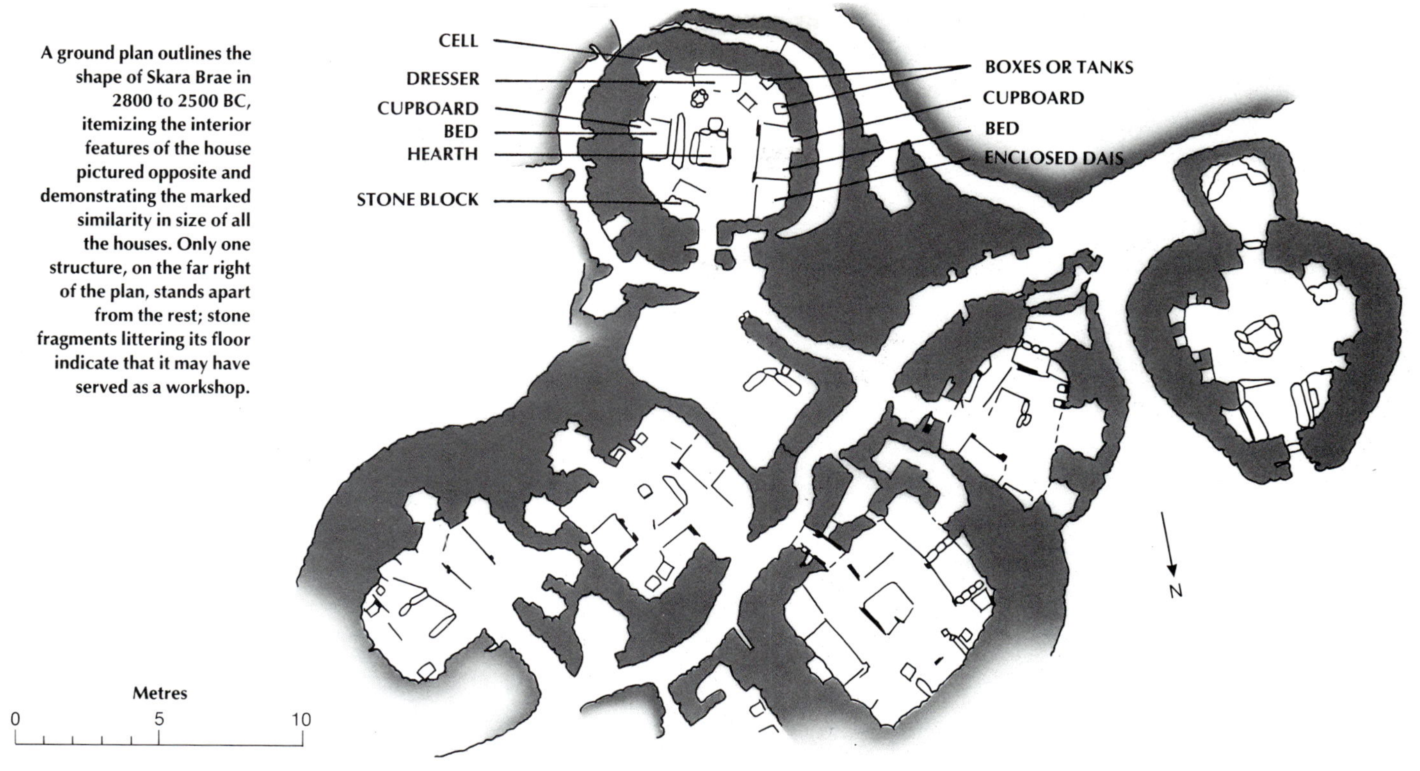

Remains of a bed box, with a cupboard set into the wall behind.

Passageway with entrance to a house on the left.

disseminated agriculture throughout the continent. But it was a hard slog. Not until the beginning of the present age did farmers burst through the tsetse belt into the more hospitable climes of southern Africa, and even then it would be centuries before the tiny settlements grew into centres comparable to those of the Middle East.

The hunter-gatherers of the Fertile Crescent had not been the only ones to feel the growing pressure for change. Around the world unsettled communities were beginning to move in the same direction, and independently were discovering new ways to control the environment about them.

In China, farming villages first arose in the north, thanks – so legend has it – to the endeavours of a mythological leader, Shen Nung. Realizing that his subjects were becoming too numerous to subsist on the meat of birds and animals, Shen Nung – who knew the taste of 100 different grasses – introduced agriculture, along with pottery making, textiles and markets. It was a time of peace and prosperity, in which people, not harming each other, had no need for defensive walls or weapons.

The myth was remarkably close to the truth. For, by about 6000 BC, farming as a new way of life had sprung up in the north around the middle stretches of the Yellow

A Saharan rock-painting from Jabbaren, an area bordering modern-day Libya and Algeria, shows early herdsmen driving two distinct types of cattle: ***Bos africanus***, with its thin, crescent-shaped horns, and ***Bos brachyceros***, with thicker and forward-pointing horns. Dated between 5500 and 2000 BC, when the Sahara was more fertile than it is today, the painting depicts a well-established pastoral life style in which herding and farming have clearly replaced the hunter-gatherer life style that had prevailed in the area since around 9000 BC.

River. It was good growing ground, free of dense forest and composed of a rich, yellow loess that retained its fertility well, yet was so fine that it could be turned with only the simplest digging sticks. The only cereal that flourished in any quantity was millet — a fuzzy-headed grass whose short growing season was ideally suited to the region's climate of cold, dry winters and wet summers.

In this rich land small settlements grew up, cultivating two species of millet — foxtail and broomcorn — and a variety of fruits and vegetables including cabbages, plums and hazelnuts. Hemp provided fibre for clothing, and the cocoons of wild silkworm moths were plundered for their sturdy filament, which, unlike flax, required no spinning to give it extra strength. There were no herds of cattle, but domesticated dogs and pigs provided a ready supply of meat, which was augmented by wild game from the surrounding countryside.

In Banpo, for example, a village set beside a tributary of the River Wei, 600 people were living in round, conical huts three to five metres in diameter, some at ground level, others with sunken floors that lay well below ground level. From the doorway, a ramp ran down to the earth floor, which the inhabitants compacted to a hard consistency, probably with the help of wooden mallets. The huts' walls were made of a mixture of mud and straw supported by six posts, with a hearth cut into the floor in the centre. Beside some of the houses, deep storage pits were dug into the ground. The whole was defined by a rough ditch, beyond which lay the village cemetery and a kiln-filled area used for pot making.

At first, life for farmers such as those at Banpo was relatively easy. The region contained only a small population, and there were plenty of new fields nearby to take over when old ones became exhausted. Once again, however, living in a settled community proved to have its drawbacks. Over-reliance on millet probably led to dietary deficiencies, and the inhabitants of Banpo were small-boned people. It was the second millennium BC before the situation was rectified with the introduction of the soya bean from Manchuria in the north. Not only did this nutritious pulse provide essential oils and protein, but it was also able to absorb atmospheric nitrogen into the soil, and the people of Banpo soon learnt the art of rotating beans with millet to maintain the fertility of the fields.

A second strand of early Chinese farming developed around the rich delta of the Yangtze River, near present-day Shanghai. It was here, in around 5000 BC, that wild rice, probably introduced from the wetlands of Southeast Asia, first became domesticated in China. The warm, moist climate was ideally suited to the cultivation of this semitropical plant, which reached maximum productivity when its roots were under water. Accordingly, where the Yangtze meandered through the plains towards the sea, the Chinese diverted some of its flow into artificial pools, or paddies. Into these submerged pastures they would then transplant the individual rice seedlings, which had first been nurtured in dry ground to a height of about 30 centimetres. When ripe, the whole plant had to be removed by teams of stooping workers. The crop was good and populations burgeoned — which was just as well, for successful rice growing required the work of many hands — eventually expanding away from the flat lands to cover the hillsides with cultivated terraces.

The drawbacks of agriculture were no less apparent in China than they were in the Middle East. In addition to the normal health hazards that came with a settled life, the Chinese suffered increased risk of infection from such water-borne parasites as liver flukes, which inhabited the paddy fields. And whatever truth lay in the Utopian peace of Shen Nung's time was soon dispelled. More than one excavated Chinese village shows signs of its inhabitants being put to violent death by marauders.

The Chinese were not the only ones in Asia to adopt farming independently of the Middle East. Since 7000 BC, the inhabitants of Mehrgarh, a small village in Pakistan's Kachhi plain, had been cultivating wheat and barley and herding livestock. And within three millenniums the settlement had become an important administrative and trading centre whose craftsmen worked in lapis lazuli, marine shells and turquoise imported from Iran. By this time, too, settlements had sprung up to the east in India's Ganges Valley, the settlers supplementing a predominantly fish-based diet with their own cultivated rice. It was not for several thousand years, however, that these advances reached southern India. Here the pastoralists of the region would set up a hut encampment on the hillsides of a valley. They constructed terraces of dry-stone walling, behind which earth was tilled and cultivated. Small, circular huts were built on each terrace, simple structures of wattle and daub, with the village extending upwards from one level to another for several hundred metres above the valley floor. The villagers grew millet as a main staple crop, which was probably shared as fodder with their large herds of zebus, or humpbacked cattle. Distinct from the wild aurochs of the Middle East and Europe, these beasts were well adapted to a tropical climate, being equipped with large surfaces that provided effective cooling – a long, narrow head, heavy dewlap, pendulous ears, long legs and a characteristic fatty hump on the back of the neck, which served both as a water reservoir and as a store of energy when food resources were scarce. The domestication of these animals became widespread throughout the subcontinent, and they were later introduced to the Middle East and eventually by way of Egypt into Africa.

By the end of the second millennium BC, agriculture had blossomed throughout India and Southeast Asia. From China, it had spread eastwards to Korea and north into Manchuria. It took a while longer for farming to reach Japan, where the rich

In a third-millennium-BC stone frieze found in the temple of the birth goddess Ninhursag, in modern-day Iraq, two limestone figures of Sumerian priests sit milking cows *(opposite page)* while tethered calves look on, snouts muzzled to prevent them suckling. In the left-hand section of the frieze, beside a cowshed, fellow priests are probably processing the milk into butter: the first *(far left)* stirs a large jar of milk, which his two colleagues strain into a pot. A fourth churns the strained cream into butter.

diversity of food resources enabled hunter-gatherers to survive quite adequately. Not until the last few centuries BC did population pressure drive them to cultivate crops — most notably rice. And there the spread stopped. Many millenniums before, Asian hunters had migrated over the northerly Bering Strait to populate the American continent. This time, however, the Americans were going to forge their own trail.

Across the Pacific, in Mesoamerica, events were taking their own course. Here there were neither the herd animals nor the stands of wild wheat and barley that had first encouraged the Middle Eastern people to settle and establish villages. Nevertheless, the region's terrain provided the gamut of growing conditions, ranging from the dry heat of the plain to the cool moistness of the mountain ranges. From among the many varieties of edible wild plants that flourished there, the Mesoamericans were able to develop their own cultivable crops.

For four millenniums Mexicans in the dry, northeastern region of the Sierra de Tamaulipas experimented with plant cultivation, until by about 3000 BC their efforts in the fields provided a quarter of their food supply. The earliest plant to be brought in hand had been squash, and it was not long before the bean had become an important staple. Like the wild cereals of the Middle East that shattered when ripe, the wild bean pod twisted into a spiral and catapulted its seeds in all directions. As with the wild cereals, there were mutant pods that failed to operate this explosive way of dispersing seed and these were picked and planted to produce a more easily harvested crop.

Together with potatoes, groundnuts, avocados and chilli peppers, these crops were cultivated in small plots or horticultural gardens near the settlement. Nevertheless, the transition to full-blown farming was, as in the Middle East, a gradual process. Wild deer and the piglike peccary were still hunted with stone-tipped wooden darts hurled with the aid of a throwing stick, mud turtles were caught, and rabbits were

killed in communal drives. When large animals were not to be found, the hunters' ovens – simple pits filled with heated stones – would be filled with such delicacies as lizards, grubs and grasshoppers.

It was 1,000 years later and 500 kilometres distant, in the Tehuacan Valley of central Mexico, that the scales first began to dip in favour of farming. As at Tamaulipas, the Tehuacans had steadily experimented with the cultivation of wild plants, in particular a tall grass that thrived around the streams that spread across the valley floor. Initially it must have seemed an unpromising source of food: the plant yielded tiny ears, about 2.5 centimetres long, bearing just a few husk-covered seeds. But in only a relatively short space of time this scanty offering was to become recognizable as modern-day maize, one of the major building blocks of American agriculture.

In the case of most other plants, it had been evolutionary backward mutations that permitted cultivation to flourish on a large scale. With the Tehuacan grass, however, nature's own efficiency came to the aid of humanity. Unlike the cereal grasses of the Middle East, those in Mexico became cross-pollinated, and as a result hybrid varieties appeared with longer ears and more numerous seeds. Planted in large numbers by Mexican farmers, these in turn released their pollen to produce further hybrids. As food production swelled so did the population, and the valley inhabitants soon found it worthwhile to make special clearings in the tangle of wild vegetation where large fields of crops could be planted on a more extensive scale. By 2000 BC the seedhead of maize had changed from something little larger than a thumbnail into a medium-sized cob five centimetres long. More importantly, the transformation had more than trebled yields to 200 kilograms of corn per hectare. It was enough to turn the balance and enable people to settle in villages as farmers.

As the good news of corn cultivation spread throughout Mesoamerica, settlements grew apace. To their diet the villagers soon added migratory ducks, the indigenous dewlapped turkey and small, hairless dogs bred especially for the table. And as in the Middle East, irrigation was used to boost production. In the southerly Oaxaca basin, for instance, the water table was in places just three metres below ground level and it was an easy matter to sink wells from which water could be drawn and channelled through ditches to reach the fields. Using such methods, it was possible to harvest two or even three crops a year.

Mesoamerica was the seedbed from which farming slowly expanded northwards through the continent. At the same time, some 2,500 kilometres south of Mexico, the inhabitants of present-day Peru were finding their own way up the farming ladder.

The terrain and climate of Peru were even more varied than those of Mexico. Although just a few degrees to the north of the equator, the high valleys of the Andes mountains provided a cool, fertile environment that abounded with game – deer, and two members of the camel family, the guanaco (an ancestor of the llama) and its thick-fleeced cousin the alpaca – and a number of edible plants, most notably the wild potato. In the river valleys that ran down from the hills, there were wild grasses and other edible plants, as well as a number of small game such as the guinea pig. On the Pacific coastline, however, the country comprised semidesert that received appreciable amounts of rain only one year in every 25.

The high Andes were the heartland of South American agriculture. It was the early hill-farmers who first began to cultivate the potato, learning how to preserve the tuber for storage. In the cold mountain air, the potatoes were left to freeze and thaw alternately, while being trampled by foot to drive out all moisture. The result was a fully dehydrated food that could be stored or ground up into flour. It was the same people who, in 3500 BC, saw the advantages of domesticating the guanaco. These sturdy, all-purpose beasts provided wool for cloth, meat that was either eaten fresh or sun-dried, sinews that were used for binding, and dung as fuel and fertilizer. But above all, the animal was capable of carrying loads of up to 60 kilograms for as much as 20 kilometres a day. Within 2,000 years the alpaca had joined the llama as a mainstay in a herding economy that stretched along the western seaboard of Peru.

All along the Pacific coastline of Ecuador and Peru, settlements based on farming were becoming established. Where soil showed signs of impoverishment, people would restore its fertility with guano – the accumulated excrement of the millions of sea birds that inhabited the rocky coastal islets. And once again, irrigation began to work its old magic. It started first in the coastal deltas, where the inhabitants dug simple ditches down which they diverted the Andean rivers' annual floods away from the sea and on to their fields. As the population grew, so the ditches snaked farther and farther towards the river source, until eventually they reached the mountains. By about 1000 BC the crude ditches had become an extensive canal network, winding around the contours of the hillsides for up to 120 kilometres. Fifteen metres above the ground, man-made aqueducts of stone and earth, up to two kilometres long, spanned whole ravines, bringing much-needed water to adjacent valleys. And these edifices meant more than just an increased food supply. They represented the first stirrings of civilization in South America.

As in the Old World, so in the New the adoption of agriculture brought as its consequence a new range of health problems. At the Dickson Mounds, in Illinois, there is

A sandstone engraving *(right)* discovered in the French Alps depicts a farmer from the third millennium BC using two yoked oxen to pull a rudimentary plough; the stylized horns of the oxen are clearly visible in the outline drawing below. The harnessing of animals for hauling came only after human beings had already learnt to exploit their skins for clothes, their bones for tools and building materials, and their flesh and milk for sustenance.

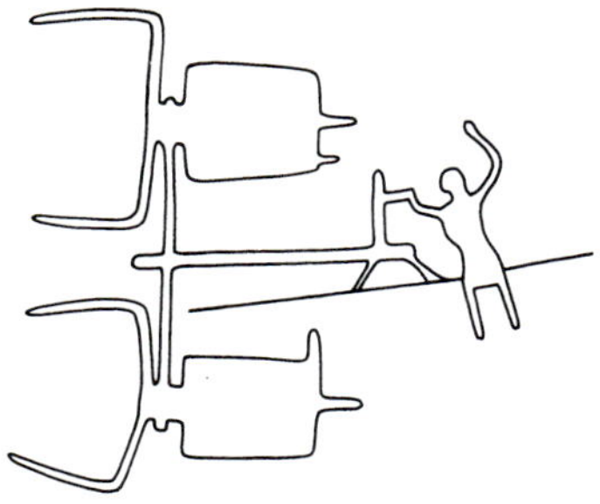

evidence that people became increasingly dependent on maize to provide their principal foodstuff. The skeletons of these people illustrate the potential hazards of such a dietary policy: marked iron deficiency and an upsurge in dental defects, both caused by lack of a proper balanced diet.

The prehistoric communities of the Georgia coast tell a similar story. Here, the remains of early farmers show that the health of women, in particular, declined after the introduction of agriculture. Whereas the female members of the community now stayed at home tending the maize crop, the menfolk still went out on occasional hunting and fishing expeditions, and these trips away from the village gave them more opportunities to eat protein-rich meat and fish.

In roughly 6,000 years, agriculture had come a long way since the mutant grains had first been noticed growing on the communal rubbish dump. It was still far from providing a panacea for the world's food problems: increased clearing of forests, overgrazing and use of the plough eventually led to soil erosion. In the Middle East, the very irrigation systems that had once proved such a boon eventually rendered much of the land useless for agriculture. This was because the fierce heat of the sun had caused the flood waters to evaporate, leaving behind deposits of salt in such quantities that plants and vegetables were unable to survive in the earth. Nor did the advent of a new way of life necessarily mean the death of the old. Until their extinction in AD 1671, the aurochs were still hunted in the primeval forests of northern Poland, and as recently as the late 19th century the situation in prehistoric Abu Hureyra was paralleled by that in southern Africa, where hunters culled migrating antelopes whose herds, estimated at 10 million, stretched from horizon to horizon. Even today, in the frozen Arctic, in the depths of the jungles of South America, and in the vast deserts of Australia and Africa, communities still pursue the hunting and gathering life style of their ancestors.

Still, once established in farming settlements there was, for most people, no turning back. The pressure of population, which had been one of the reasons why humans had taken to farming, was now an overwhelming obstacle to its rejection. The only way was forwards. Humanity had learnt to control nature: it was now faced with a new challenge – that of controlling itself.

For agriculture had done more than simply provide an alternative means of survival: it had radically changed the whole fabric of human society. With enough to eat and more time on their hands, settlers could now turn their attention to matters other than food production. Those who had the ability found it possible to earn a living from selling the products they made: tools, weapons, grave goods, cooking utensils and personal adornments. As trade boomed, so increasingly people came to live off each other rather than off the land. And, inevitably, some – those who were better craftsmen than others, perhaps, or whose lands were more productive – became wealthier than their fellows. It was a wealth that not only gave them the status of employers, but also could be passed down through the family line. The elementary beginnings of a hereditary class system had emerged.

More than either wealth or opportunity, the new social hierarchy meant authority. Someone had to oversee the construction and maintenance of irrigation systems, to coordinate the settlement's defensive arrangements, to administer religion and to regulate and record trade. These new features would emerge as the distinctive hallmarks of society's progression towards urban life.

A NEW DESIGN FOR LIVING

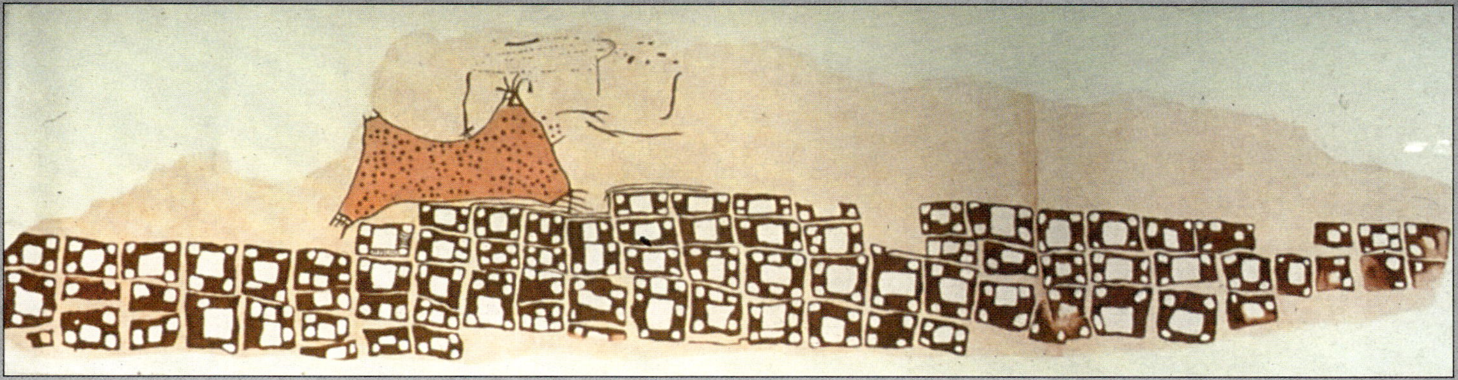

More than 8,000 years ago, a bustling, thriving, trading centre known to us as Çatal Hüyük, the "forked mound", stood overlooking the sweeping plain of what is now south-central Turkey. A copy of a wall painting *(above)* found in one of the town's shrines depicts Çatal Hüyük as it looked in around 6700 BC. Close-packed rectangular houses rise in graded terraces, while behind them the twin-coned volcano of Hasan Dag, 135 kilometres to the east, belches fire, smoke and molten lava.

This busy community of several thousand souls grew wheat, barley and peas in the surrounding fields. From the hills and mountains came almonds, pistachio nuts and fruit, and for their meat the townspeople not only hunted the herds of wild pigs, red deer and aurochs that roamed the open flat lands, but also ate the cattle and sheep that they themselves had domesticated. The town's prosperity seems to have stemmed in part from the region's plentiful supplies of obsidian – a shiny black volcanic glass used to make knives, mirrors and jewellery – which they excavated at Hasan Dag and traded for items such as decorative Mediterranean seashells and good quality Syrian flint. Çatal Hüyük may also have been a religious centre. Only about a half of one of the town's 13 hectares has been excavated, and in this area alone one building in three seems to have been a shrine.

For all their piety, many of the town's inhabitants clearly failed to ward off misfortune. Their remains show signs of the skull-thickening caused by malaria, of injuries sustained in rooftop falls and of head wounds caused by blunt instruments. For the most part, though, the inhabitants of this fledgling urban community enjoyed a level of security and sophistication unattainable elsewhere, as depicted in the illustrations on the following pages, based on archaeological finds made at the site.

A TOWN WITHOUT STREETS

The people of Çatal Hüyük built their houses so close together that the rooftops had to serve as the town's thoroughfares, with wooden ladders providing a means of getting from one layer to another. Entry to each house could only be obtained through a small door on the second storey. With no access at ground level, the town presented a solid bulwark of blind walls to floods or attackers — a feature that made Çatal Hüyük a far safer place to live than most settlements at that time. When it rained, gullies between the houses drained the water off the roofs; down below, the sites of derelict houses served as rubbish tips and latrines, over which the townspeople sprinkled layers of ash to stifle the smell. Though the buildings all looked somewhat uniform, each household had its own distinctive clay seal *(below)*, for marking clothing or bags of grain, or even for decorating its owners' skins.

A cutaway floor reveals a skeleton wrapped in a coiled basket and buried beneath the spot where the mother and children of this house would sleep. When someone died, relatives left the corpse on a platform outside the town, where vultures and insects picked the bones clean. Then they took the skeleton home for burial, perhaps during a spring festival in which townsfolk mended walls and roofs.

Pictured here is a typical family home, a single room of about six by 4.5 metres. A ladder descends from the rooftop entrance, hung with pig bladders full of drinking water. Two oak rafters support closely packed crossbeams, over which lie woven grass mats, bundles of reeds and a hard outer layer of dried mud. In the foreground, a low door in the inner wall leads to a storeroom housing clay caskets and jars. Under the stairs is the kitchen area, with its open hearth, an oven in the wall and a variety of utensils such as a burnished cooking pot and a bone spatula and two-pronged fork *(below)*.

A DWELLING FOR THE DEITY

Symbol of male potency, the plaster head of a bull, flanked on either side by a row of plaster female breasts, thrusts out of the wall of this Çatal Hüyük shrine. A painting on the left-hand wall depicts giant vultures with outspread wings picking a headless corpse clean prior to burial. On the floor, skulls of the dead sit in mute homage to the shrine's deity. Devotional objects excavated at Çatal Hüyük include a ceremonial flint knife, with its bone handle carved in the form of a snake, and a marble statuette of what is thought to represent a mother goddess figure *(below)*.

ARCHITECTS OF A NEW SOCIETY

4

Without the sweet water from their spring to refresh them, the inhabitants of Jericho could never have found the strength required to build the wall that encircled their town by 8000 BC. The digging of a broad defensive ditch and the gathering and piling up of stones into a rampart was thirsty, painful work. The construction of a massive watchtower, rising some 8.5 metres from the ground, would be an even more arduous task; there would be angry words, accidents, perhaps even deaths. But these new defences were a vital necessity. Their settlement, near the northern end of the Dead Sea, was an oasis in the arid land that later generations would call Palestine. Their successors would make war over the ownership of the region and its precious water sources for another 10,000 years or more.

To nomadic bands emerging from the southern deserts, or migrants descending from the northern heights, the huddled oasis settlement of round, mud-brick houses must have exuded an enviable prosperity. In 8000 BC, Jericho's fields, nourished by the stream that flowed from the ever-generous spring, grew valuable emmer wheat in abundance; the surrounding wilderness, with its natural population of gazelles and ibexes, provided a seemingly unending supply of fresh meat. As many as 1,000 people – an astonishing prospect in a landscape that was still populated mainly by either small, struggling settlements or wandering tribes of nomads – dwelt here together, living off the fat of the land.

But Jericho's Neolithic inhabitants could do more than simply provide enough food for themselves: they possessed a rich surplus of material goods which they were able to send away in bags or baskets on the backs of itinerant traders, to be bartered in exchange either for materials they could not find on their own home ground, or for some useful tool or utensil crafted in a manner with which they were not familiar. Prosperity brought with it the opportunity for them to stand back and consider their shared situation; they were organized enough, and articulate enough, to cooperate in enterprises that would benefit the common good. Faced with an expanding population, they had agreed upon a means of channelling the stream that ran below the settlement, allowing them to water the fields and thus increase the wheat crop. One successful engineering project may have led to another. They had learnt how to pool ideas and effort, how to make a plan, carry it out and get results, and how to build a wall to protect themselves against attack.

In the end, though, their ramparts were not enough to save them. For generations, perhaps for many centuries, the Jerichoans carefully maintained their defences in good order, but at some point during the eighth millennium BC their community disappeared. The cause may have been some natural disaster, or an invading tribe that swept the inhabitants away and left no trace of their attack. Yet a site with so many obvious natural advantages would not remain empty for ever. Within a few centuries,

An inhabitant of Jericho, a city northwest of the Dead Sea that flourished around 10,000 years ago, is immortalized in this lifelike head, modelled in plaster over a human skull. A number of such heads have been found buried beneath house floors – with cowrie shells for eyes and realistic facial features, they appear to represent particular individuals. The heads might have been meant simply as family mementos, or they may have played a part in a cult of ancestor worship.

The world's first settled societies, which came into being sometime between 10,000 and 2500 BC, were concentrated in four main areas: the Near East, the Nile Valley, the Indus Valley and the North China plain. Sited on or near rivers within a belt of largely equable climate, between latitudes 24 and 38, these clusters of towns and cities owed their origins and continued survival to an abundant agricultural supply from the surrounding region.

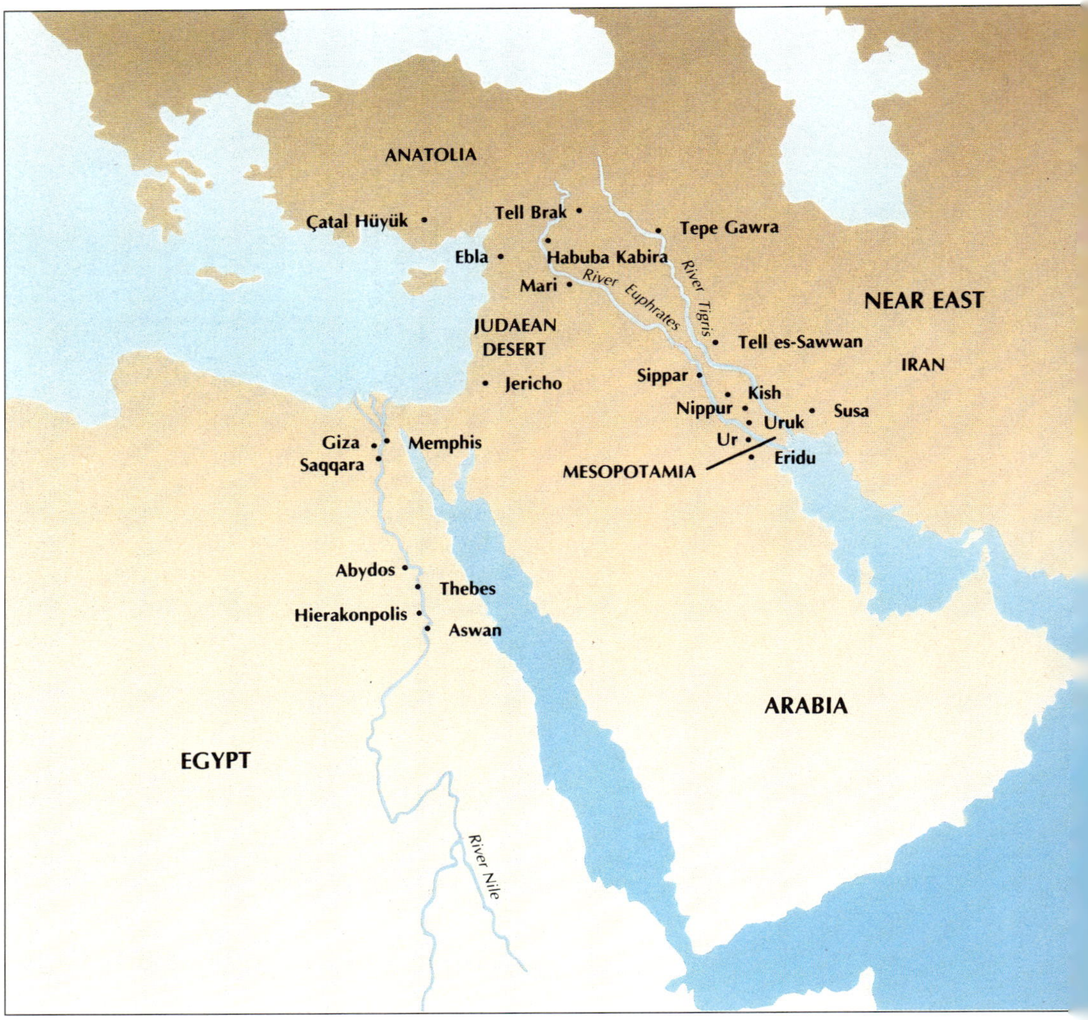

possibly even sooner, a new population, with no apparent links of kinship or culture to their predecessors, had established a new settlement on the mound that now concealed beneath it the ruins of the old.

For thousands of years Jericho's spring would nourish the crops of a succession of different populations who were in the process of moving down from the mountains or emerging from the desert. Some would be better farmers than their predecessors, others less advanced. Some would be skilled at building solid houses, while others — centuries after them — would camp in roughly dug pits, roofed only with animal skins stretched on frames. Some would be gifted potters, whose bowls and drinking cups were as beautiful as they were functional; their successors might rely only on the crudest, most primitive ware. But each group would learn new skills in its own time, and build upon that knowledge. Then some disaster would occur — a plague, an earthquake, or a shrieking horde of raiders — and these people too would be gone. Sand would cover their ruined houses, mud would clog the irrigation channel, and the ancient mound would stand empty until new arrivals came to build on it again.

This pattern would repeat itself, not only in the neighbourhood of Jericho, which was located on a crossroads of many migratory routes, but on innumerable other sites

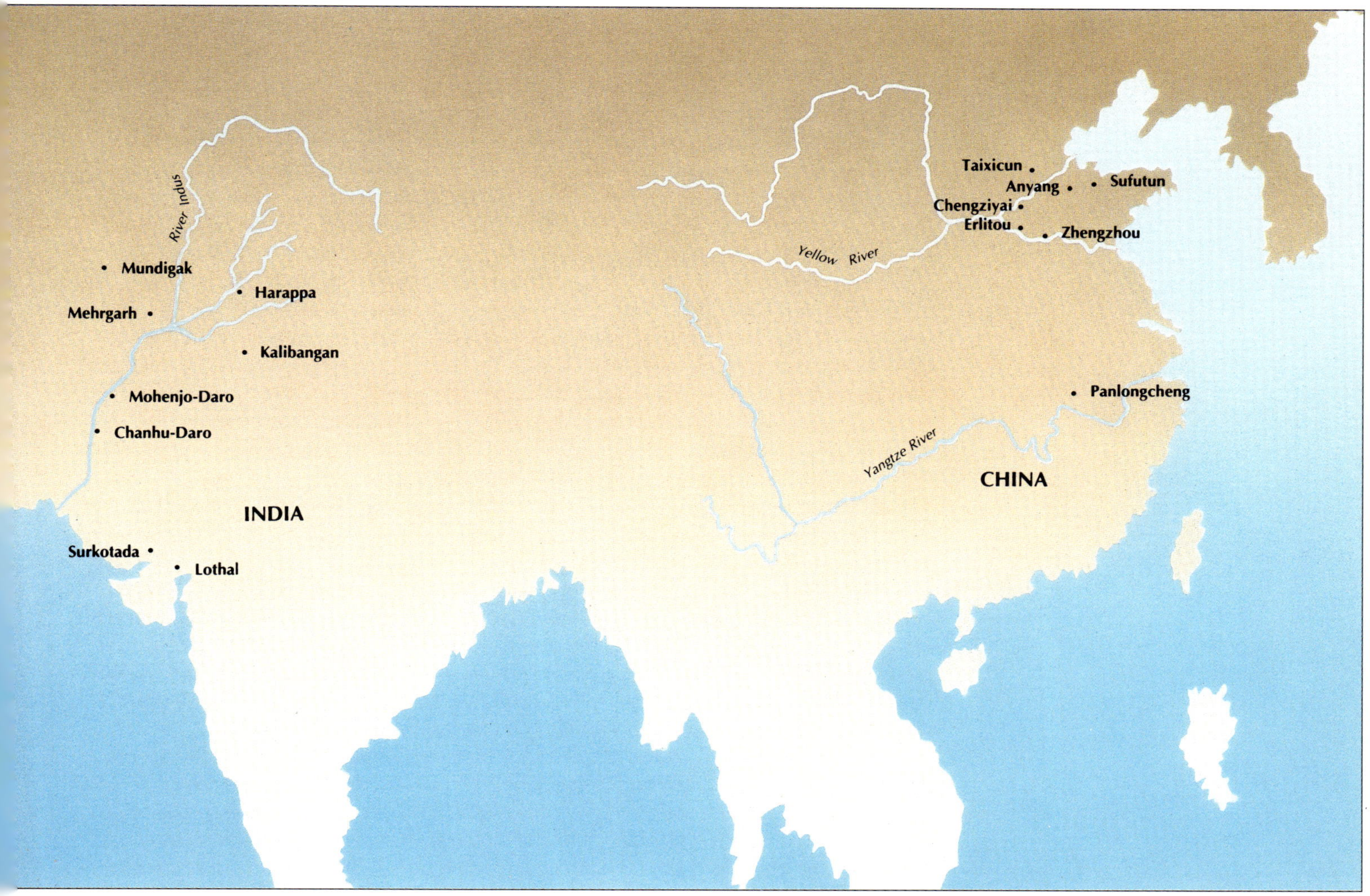

in every inhabited region of the world. In the Judaean desert, as in China, the Indus Valley or the delta of the Nile, civilization was not destined for an easy birth. There were times when it seemed as if every step forwards was followed by two steps backwards. Humankind, in the process of waking to its possibilities, did not march along a single, straight road, but threaded its way through a succession of mazes composed of switchbacks, dead ends, and endlessly proliferating branches. But, despite the many obstacles in its way, still it progressed.

The first wall-builders of Jericho, laboriously hoisting their stones into position, had been merely a single cell in the revolution that was taking place throughout the inhabited world between the 10th and third millenniums BC. During this period the people of the earth began to reap the rewards of their own intellectual and technological advances. They began to take control of their environments, producing their own food, creating new tools to ease and speed their labours. They discovered how to live and work together in large communities, and began to trade – not only goods, but knowledge – between one settlement and another. At the beginning of the time span, human life was lived beneath the constantly looming shadow of famine and disease. By its end, communities were still just as likely as before to be suddenly wiped out by starvation or sickness, or the new scourge of warfare, but culturally and socially they had progressed far beyond their predecessors. Civilization had begun.

The engine that drove these movements was the community. As women and men learnt more about the world, the words and phrases they used to exchange ideas and information grew correspondingly more complex. The past could be remembered, the future planned for. People who were descended from the same tribal ancestors, or else had settled close together, understood each other; they saw the world, given shape by their words, in a similar way. Those from another stock, or who migrated far afield and kept no contact with their kindred, employed different sounds with different meanings. Even those tongues that evolved from a common root soon diverged until all mutual comprehension was lost. The paths that these ancestral languages travelled were as varied as the paths their speakers took to populate the earth, and as subject to speculation. But words, at least those uttered in the time before writing was invented, disappeared on the wind; unlike stone shelters, flint tools, bone ornaments and even the action of fire upon wood, they left no traces.

Still, something did survive to indicate to later generations the ways in which these early men and women thought and acted. Their sites of worship remained. Massive boulders were dragged many days' journey from the places where they had been quarried, to be set up in circles or aligned in rows that strode across the landscapes of Britain or Brittany. Other later edifices, carved with the sombre faces of the gods, stood vigil on remote islands in the Pacific Ocean, or vanished deep beneath the green profusion of a Central American jungle. In the east of Ireland, a richly decorated burial chamber was carefully positioned so that it would receive within its depths a single, hopeful shaft of sunlight on the darkest day of winter.

In Britain, stone slabs, twice as tall as any man, were planted in pits in the ground, two or three or 60 of them together, spaced at regular intervals. Those who had set them in place chose configurations that might have served as a massive sundial; others could have been used to chart the slow, seasonally changing processions of stars in the night sky. Whatever their function – the honouring of invisible powers, the measurement of time, the commemoration of some significant

Occurring naturally in lumps and seams as well as in ore, copper was the first metal to be put to use by humans. As early as 9500 BC smiths in southeast Europe and western Asia were hammering the metal cold to produce ornaments, tools and weapons. Later they found that heating the metal made it stronger and more malleable, and around 6000 BC they learnt to extract copper from ore by applying intense heat. Still greater temperatures were needed to melt copper so that it could be poured into clay or stone moulds to cast tools such as knives and axes. Further advances later enabled smiths to work with bronze – a mixture of copper and tin – iron, and, by 150 BC, steel. Examples of their increasingly sophisticated work appear on this page and overleaf.

WORKING WITH METAL

A hammered copper blade from Syria is flanked by two delicate, cast toggle pins, one with the head of an ibex or gazelle, which were probably used for fastening cloaks.

In a frieze from Al 'Ubaid in Mesopotamia, a lion-headed eagle representing the god of storms grips the tails of two stags. Dating from around 2700 BC, the panel of carved wood is covered with beaten sheet copper, fixed by copper nails. The stags' antlers, also copper, are soldered in their sockets with lead.

event or other — these relics spoke of shared beliefs and common purposes. So too did the images that people painted on rocks, that they carved and kept in niches in the walls of their houses, that they tucked lovingly into the burial pits along with their dead. The deities they worshipped varied in form and gender. Some were given the shape of beasts while others remained faceless entities, too potent to contemplate directly, and represented only in the form of symbols.

The newcomers to seventh-millennium Jericho arrived on the mound many years after their predecessors' disappearance. The work of wind and weather, the erosion of the soil, had covered the traces of this earlier community so completely that its successors were barely able to make out the remains of the earlier walls and houses. These new people brought with them a uniform style of architecture, and a conformity in their domestic arrangements that suggested that they were not nomads, but veterans of a settled life in some unknown homeland. They built their large rectangular houses from hand-moulded, sun-dried mud bricks that were the shape of flattened loaves; these were held together by mortar, at that time a new invention, most probably made by mixing heated limestone with sand and water. On the bricks' upper surface the Jerichoan brickmakers would impress a row of thumbprints in a herringbone pattern. When construction was under way, the builders would pour the mortar into these indentations until it began to overflow, at which point it was ready for the next brick to be placed on top.

Prudent housekeepers, the Jerichoans covered the clay floors of their houses with tough, hard-wearing lime plaster, which came in either cream or red. Those who would be responsible for the housework evidently had their say in the design too; the plaster and the wall joined together not at right angles, where dust and dirt would

collect in the crack, but in a smooth upwards curve that would allow dust no place to hide from their brooms or leather mops. Outside, these house-proud types positioned vats against their walls to collect precious rainwater, and did their cooking in a central courtyard, grinding einkorn wheat into flour in querns, and eating plenty of meat. Like the very first occupants of Jericho they relished the gazelles brought in by the hunters, but also ate the flesh of sheep and goats that were either wild or in the earliest stages of domestication.

Naturally, the fertility of their fields, and of the beasts they either reared or hunted, were matters of profound importance to the wellbeing of the community. In hopes of cajoling, or controlling, the forces of nature, they shaped small animals out of clay, and formed female figurines with generous curves – images of fecundity and the mysteries of life's creation. Certain buildings within their settlement seem to have been set apart for ceremonial purposes; one of these had two rounded alcoves in which sacred objects might have been lodged, and a central basin that may have received the blood of ritual sacrifices.

The dead were accorded their own kind of deference. Their bodies were taken from their houses and set apart until the flesh was gone – decayed or stripped away by animal predators. Then the skulls were carefully brought home again, and buried under the plaster floor, allowing the house's living occupants to pursue their daily tasks in comforting proximity to their departed kin.

These practices were by no means unique to Jericho. Skulls found buried in clusters under similarly decorated rooms in smaller settlements in what is now Syria suggest that the Jerichoans may have been members of some larger grouping, to

A Mesopotamian charioteer drives a team of onagri, or wild asses. This statuette from Tell Agrab, made around 2700 BC, is the first known example of the "lost-wax" method: a clay mould was made over a wax model, the wax was heated and drained off, and molten copper poured into the hollow mould.

which they were allied not only in their shared spiritual beliefs, but in the language that they spoke, the tools that they used, and the customs that they lived by.

Increasingly, as populations expanded and settlements spread, their members established peaceable contacts for the purposes of trade. Surplus foodstuffs constituted only a small part of what they traded. Craftsmen everywhere were prepared to pay good prices for the raw materials they needed for their trade, or the substances that would give an attractive colour to their finished wares. Beidha, for instance, a small village in Jordan, based a thriving trade on its abundant supplies of the red iron ore hematite, which gave pottery a handsome red colour and which could in addition be used to make the rouge that women rubbed on their cheeks, or applied to their lips with tiny spatulas formed out of carved bone. Small settlements in Iran traded lapis lazuli and carnelian for jewellery, and soapstone, which was soft enough to be easily carved into smooth, pleasing shapes.

Jericho, enjoying easy access to the Dead Sea's salt deposits just 12 kilometres away from the settlement, found itself endowed with a highly prized commodity: people craved salt in their diet, and discovered how efficiently it was able to preserve their surplus foodstuffs from decay. Relayed by means of a chain of middlemen that might extend for hundreds and even thousands of kilometres, these goods were exchanged for cowrie shells gathered by dwellers on the Mediterranean seashore, for turquoise from the Sinai peninsula, or for the hard, glassy, black substance known as obsidian, which was sharp enough for knife blades and shiny enough for mirrors, and which was found on the slopes of extinct or dormant volcanoes. In the eyes of the Jerichoans, obsidian was also invested with an additional layer of desirability because of its having originated on the distant, mysterious plateau of central Anatolia, some 800 kilometres to the northwest.

A large part of the Anatolian obsidian trade was conducted by the residents of Çatal Hüyük. Unlike Jericho, Çatal Hüyük suffered no disruptive cycles of abandonment, destruction, repopulation and rebuilding. A single people occupied the site, with no indications of significant disturbance, for at least 1,500 years. When they eventually departed, it was only to build themselves a new town on a site across the river, where they prospered for another 800 years before disappearing – again, without any signs of violence or conquest – from the scene.

Enjoying as they did the benefits of continuity, prosperity and tranquillity, the inhabitants of Çatal Hüyük also had the freedom which allowed them to develop an extraordinary repertoire of crafts and talents. Their town was a wonder for its age – an apparently affluent community of scrupulously clean, well-furnished households and elaborately decorated shrines.

Religion, or its outward trappings, formed the constant backdrop to Çatal Hüyük's domestic life. In at least one part of the town – indeed, by some reckonings this well-ordered and apparently populous settlement would have to qualify as an embryonic city – one in every three buildings served as some kind of ceremonial chamber or shrine. Mother goddesses, vibrantly painted or sculpted in dramatic relief, gave birth to divine daughters while resting on the shoulders of sacred leopards; quintessential images of gender – masculine horned beasts and women's breasts – projected from walls or benches. Over the course of the settlement's long life they demonstrated an apparent shift in values: the horned skulls of male animals, sacred to the hunter, gradually vanished from the holy chambers, first outnumbered and finally replaced by images of the life-giving female, whose fertility mirrored that of the earth itself. The hunter surrendered his primacy to the farmer tending the community's fields.

To glorify their gods, furnish the burials of their dead, and enhance their own domestic lives, the people of Çatal Hüyük sought far and wide for the materials their craftsmen needed. Timber for building – oak and juniper wood – came from distant, wooded slopes, where it was cut into logs and floated down the rivers to the plain. Limestone, alabaster and fine white marble came from quarries many days' journey from the town. Shells of sea creatures – cowries, dentaliums, whelks – came all the way from the Mediterranean coasts. Traders, climbing the foothills and negotiating the mountain passes, kept a sharp eye out for signs of the bright minerals and metal ores that painters used to make their pigments: blue-green malachite, sky-coloured azurite, brown and yellow ochres, deep red cinnabar.

The work of its craftsmen testified to the fact that Çatal Hüyük was a town of specialists; no footsore hunter or hard-pressed subsistence farmer would have been able to spare the time, the energy or the daylight required to achieve such levels of skill. Every kind of stone was ground, polished, pierced or carved to make heavy-duty sickle blades and axes, pestles and mortars, adzes or chisels for fine woodwork, palettes for paints and cosmetics, and snake-handled ceremonial daggers realistically adorned with serpents' eyes and stippled scales. Carpenters knew how to construct squared roof beams, doorposts and sturdy ladders, as well as wooden tableware of advanced design, from great meat platters with carved handles to sauce boats and tiny egg cups. Weavers used fine rushes and marsh grass to make floor coverings, understood how to turn plants into dyestuffs, and used clay stamps incised with different geometric designs to print colour on to cloth. The vivid hues and patterns they worked into their textiles were copied by painters on to the walls of shrines and houses; the same types of design would be preserved and reproduced by the women of the region for 8,000 years to come.

The craftsmen of Çatal Hüyük also possessed another extremely valuable skill – the art of metallurgy. They knew how to apply heat to lead and copper, and how they could then hammer the softened metal into various shapes, turning out a range of beads and ornamental cylinders, which might have been strung on pendants or threaded on to the fringes of clothing. Whilst they were among the earliest practitioners of the craft, however, they were not necessarily its pioneers. The discovery came first, some time between 7000 and 5000 BC, among settlements scattered throughout western Asia and southeastern Europe; it would be another 2,000 years or more before it penetrated to northern Africa and the Far East.

The knowledge that heat could be used to draw metals from particular rocks, and that these metals could be manipulated into different shapes and conditions, moved gradually along the expanding trade routes. The man who carried a bundle of copper beads to exchange for other desirable objects might travel no more than 50 kilometres from his base, but the goods themselves could be handed on from one trader to another, to end up halfway across a continent. Inevitably, information about their making travelled with them. An itinerant metalworker, brave and enterprising enough to leave his people, might speed up the transmission of his skills – or might learn some new technique himself and broadcast it more widely. Less peaceably, invaders possessed of the technology introduced it to the territories they conquered.

Naturally occurring nuggets of copper ore could be simply picked up from the ground in areas such as Iran and Turkey, where they were to be found in abundance.

But the working of these raw materials was only able to proceed when craftsmen began to understand the mysteries of applying and controlling intense heat. Both the metalworker and the potter were faced with a similar set of technical problems: the construction of ovens able to withstand powerful temperatures, and the need for moulds and tools that could be used without risking harm to the artefacts or injury to their maker. Kiln-fired pottery predated metalworking, but it is not inconceivable that both craftsmen actually took their lead from the baker, who would have employed a primitive oven to turn ground grain into bread.

Gradually, the metalworkers discovered more efficient ways of performing their craft. From merely hammering out copper nuggets into random, brittle shapes, they learnt how to make the material more malleable through the application of heat, originally by simply placing it in an open fire until the nugget became soft enough to work. Then, as nuggets of copper became increasingly difficult to find, they also learnt to extract the metal from more readily available ores such as malachite and azurite by smelting them in an enclosed oven.

It would take another thousand years before they devised the more complicated process of casting copper in open moulds. The technique required the development of an oven capable of generating and withstanding the fierce temperature of 1,083 degrees Celsius, the point at which copper itself – and not just the impurities it contained – would liquefy. The molten metal was then transferred into carefully shaped moulds of clay or stone and allowed to cool into the desired form, before it was hammered into a finished product. Gradually, the techniques for making the moulds grew more elaborate; one modification the craftsmen learnt, for instance, was that a charcoal cone could be inserted into a mould to produce a hole for a shaft or a handle in a blade. By the fourth millennium BC, metalworkers had mastered their art so successfully that they were able to produce artefacts such as the exquisite little specimen found in a tomb at Tepe Gawra, in what today is northern Iraq: several pieces of electrum – a natural alloy of gold and silver – expertly welded together to form a smooth, silver-coloured wolf's head, three centimetres long and complete with upright ears and curved, bared fangs.

As more and more people acquired the techniques necessary to this new skill, so metals became an increasingly valuable commodity throughout the trading networks. The nuggets of natural ores that simply lay exposed on the surface of the ground would no longer suffice to meet the growing demand, and men learnt to dig into the earth in the neighbourhood where these substances were found. Mines near Eilat, for instance, at the northernmost tip of the Gulf of Aqaba, consisted of shallow shafts leading down to a network of narrow subterranean galleries, carefully dug to follow the course of the ore-bearing strata embedded in the rock. The miners who worked them must presumably have been either very young children, or members of some unusually small-boned tribe: the tunnels would have been impassable even for the slightest full-grown adult of that time.

For those with the resources available on the spot, or with the wealth to acquire them through trade, metal tools and weapons now became a realistic possibility. Yet many generations would come and go before metalworkers eventually perfected the technique of alloying copper with tin to make the far tougher, more durable metal known as bronze. This new and much sought-after material would emerge first in the Near East, some time between 4000 and 3000 BC; it would take another thousand years or so to penetrate to the farthest reaches of Europe and Asia. When it did so it

Polished to a lustrous shine, beads of precious stone are strung in colourful sequence. The upper necklace, from Ur in southern Mesopotamia, includes carnelian, chalcedony and lapis lazuli. In the lower one, dating from around 5500 BC and found at Tell Arpachiyah, in northern Mesopotamia, lozenges of black obsidian alternate with cowrie shells.

would transform the face both of warfare and of society in general, setting apart the powerful elites, with their bronze weapons and armour, from the common mass of people armed only with bows and arrows, maces, or axes of stone.

Meanwhile, as the early pioneers of metalwork were stoking their ovens and designing their moulds, revolutionary social developments were taking place in western Asia, in the fertile alluvial plain that lay between the rivers Tigris and Euphrates.

The society in Sumer known as the 'Ubaid — flourishing between the fifth and fourth millenniums BC in what today is southern Iraq — was cursed with as many handicaps as it was blessed with advantages. The region's soil was rich and easy to cultivate, but the climate was arid, with long, hot, rainless summers. Its southern reaches, where the rivers met the Persian Gulf, were a zone of lagoons and marshes, full of fish and game, hospitable to the useful date palm, but salty and subject to tidal floods. Farther upstream, farmers had a different problem — how to bring water to their parched fields. The solution they found was to enlist the cooperation of every single farmer who cultivated land along the banks of the rivers and their tributaries.

Working together, the farmers constructed an elaborate system of irrigation channels, and organized themselves to ensure that the network they had created would be properly maintained. The benefits they gained in the process went far beyond the carefully orchestrated flooding of their ditches: their successful collective efforts laid the groundwork for other forms of collaboration. These might have included such material enterprises as the construction of defensive walls and communal granaries, or more spiritual endeavours, such as the building of a great temple in honour of the gods who controlled their harvests.

As communities grew and prospered, new, grander types of architecture began to appear. As early as the sixth millennium BC, there were houses with up to 14 different rooms at Tell es-Sawwan, an evidently affluent village in central Mesopotamia, bordered by the River Tigris and protected both by a buttressed wall and a three-metre-deep defensive ditch. Here, as in Çatal Hüyük, the people buried their dead beneath the floors of their houses, and the grave goods placed with them display a high standard of craftsmanship in materials which must have been expensively imported from far afield: figurines of alabaster, and beads of turquoise and carnelian, probably transported from Iran.

At the same time, a striking new type of luxury pottery made its first appearance in northern Mesopotamia. Named after the settlement of Tell Halaf in northern Syria, the new style was characterized by bold, linear patterns, checks and zigzags mixed with floral and fish-scale motifs, and the use of a wide range of reds, browns and blacks in the colour scheme. The widespread distribution of the pottery – it has been found as far as 250 kilometres away on the Turkish Mediterranean coast – shows that although its fragility must have made it awkward to transport, it fetched a high enough return to make long-distance trading worthwhile.

The people of 'Ubaid Sumer lacked many natural resources. They had plenty of reeds and mud but little stone, and none of the minerals needed for metalwork. But geography was in their favour; men and merchandise could move easily across their flat landscape. Trade routes crossed the increasingly well-settled Mesopotamian plain, linking the mountains of the Zagros on its eastern edge with the villages that hugged the fringes of the Arabian Desert to the west, or traversing the wilderness to reach the Mediterranean coast. Impelled by necessity, they looked to commerce as another form of farming, scattering merchants just as farmers scattered seed. The pottery vessels their traders used, whether to carry goods for sale or to hold their own provisions, could be found as far apart as the northern reaches of present-day Iraq and the eastern shores of the Arabian peninsula.

The movement of trade goods between far-flung mines and workshops, farms and fisheries, inevitably led to improvements in the means of transport. The domestication of beasts of burden, such as the donkey or the ox, allowed a trader to carry more wares than he could possibly hoist on his own shoulders, and to take them farther. It was a short step from the use of pack animals to the invention of the cart. It may have been potters who, in the fourth millennium BC, first employed a wheel to speed the working of their clay; if so, it was not long before someone thought to turn the circle on its side to provide a means of locomotion.

The people who lived along the lower reaches of the Tigris and the Euphrates were certainly seafarers. Simple rafts had been used for thousands of years, but ships with sharp prows and sails first appeared in Mesopotamia during the 'Ubaid period. After they had negotiated the lagoons and marshes at the rivers' mouths, it would have

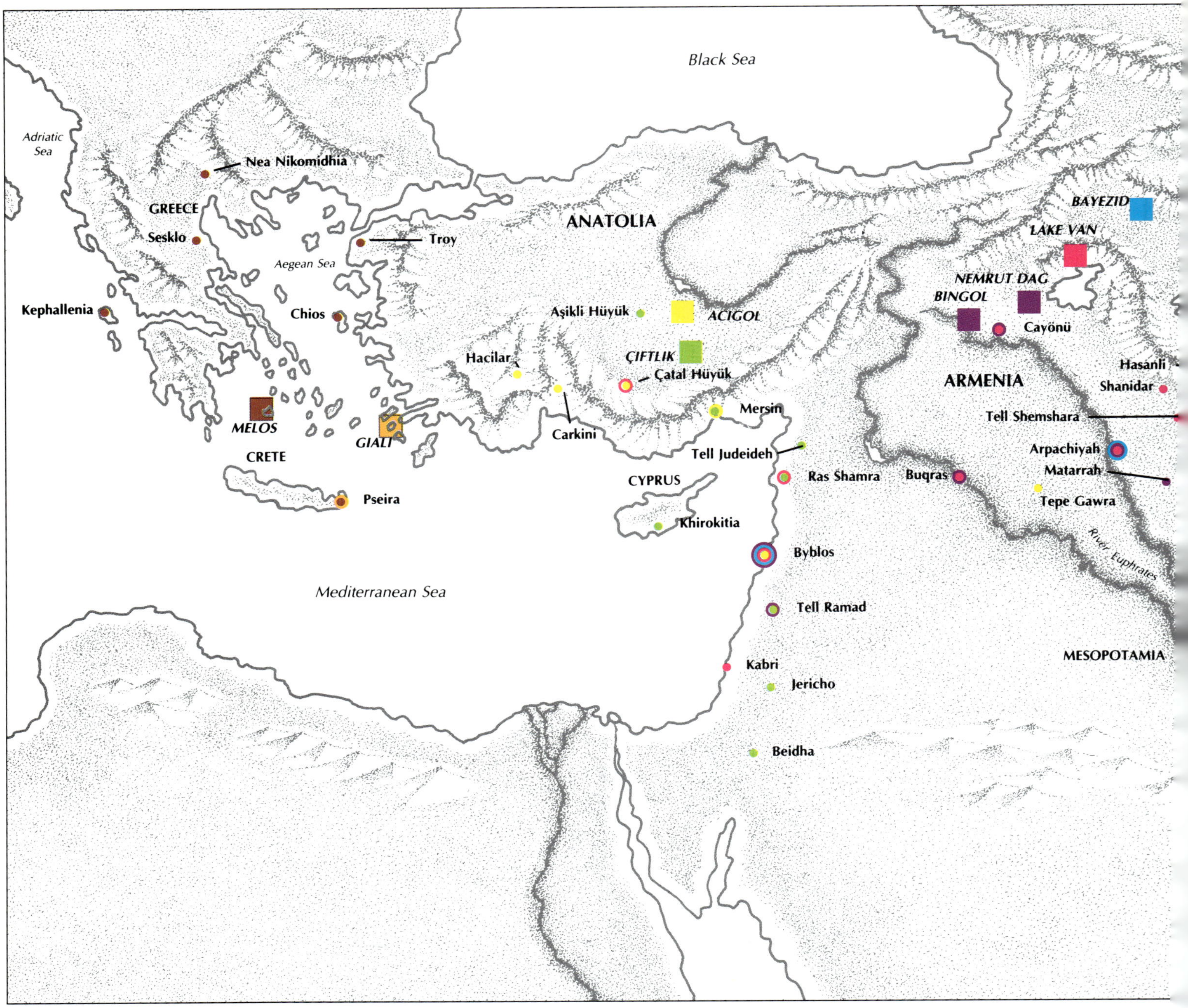

In the period between 7000 and 5000 BC, towns and settlements across the Middle East were linked by trade routes along which merchants transported the prized commodity of obsidian, a hard black volcanic glass formed by the rapid solidification of molten lava and used to make knives and arrowheads before the introduction of metal. The presence in these artefacts of varying amounts of elements like iron and sodium has enabled archaeologists to pinpoint the source of the stone. The map shows the location of obsidian-producing areas *(coloured squares)* and which nearby settlements used each source's output *(dots or circles coloured to match source)*. The obsidian from two places in eastern Turkey – Bingol and Nemrut Dag – is indistinguishable, so both are in the same colour.

been an easy enough task for mariners to hug the Arabian coast as they plied a course down the Persian Gulf. 'Ubaid adventurers undertook these voyages in search of pearls or shells, or as straightforward commercial fishing expeditions. The traces of their Arabian camp sites suggest they visited the same sites at regular intervals, and stayed ashore long enough to give them time to dry and preserve their catch, for eventual sale in landlocked markets.

Dried fish, pearls and decorative shells would all find ready purchasers in Eridu, which the Sumerians would later refer to as "the oldest town in the world". Built on

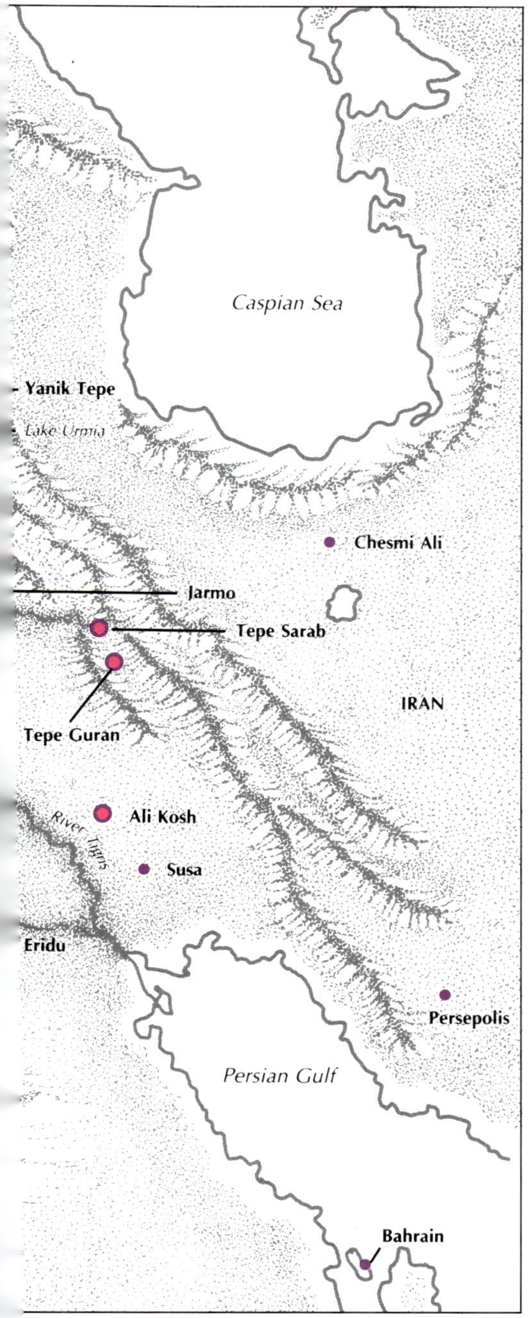

the extreme southwestern edge of the Mesopotamian alluvial plain, Eridu stood on an apparently inhospitable site, constantly menaced by encroaching sand dunes. Nevertheless it housed a thriving community of specialists, including among them not only such craftsmen as potters, weavers, bakers and brewers, but also those much sought-after experts in the all-important mystery of how to satisfy the gods: the priests and priestesses. For Eridu was a pious place, home of the water god Enki, who was believed to preside over the canals — fed from the Euphrates some 20 kilometres distant — that were essential to ensuring the city's continued survival. At the heart of Eridu stood a temple, which would survive, carefully preserved within the foundations of its larger successors, for centuries. Within these sacred precincts animals were sacrificed and their blood poured over the altar as a libation. Worshippers brought the god offerings of food and drink; Enki, it seemed, enjoyed the same refreshments that his mortal subjects would have relished — bread, beef, lamb and beer. The contributions also included sea perch, and other saltwater fish — further evidence that the 'Ubaid people were seafarers.

Outside the precincts of the temple lay a large cemetery where the dead lay along with their collections of grave goods: pottery vessels, clothing adorned with bands of coloured beads, even clay figurines which represented the gods themselves, personified as male and female entities wearing bitumen headdresses on their lizard-shaped skulls. Each grave consisted of a rectangular shaft, lined with mud bricks. Some form of tomb marker or headstone must have been used to indicate the whereabouts of each individual grave and the identity of its occupant, for the tombs were often opened — in an orderly manner, with no guesswork — to allow for a second interment. In one case the secondary burial was not of a spouse or a child, but of a dog. Among the 'Ubaid dead there were few signs of any distinction on the grounds of wealth or social rank — no luxuriously furnished tombs or expensive parting gifts that would put humbler corpses to shame.

Yet this apparent egalitarianism failed to survive. Communities grew more populous and more diverse, and the interests of farmers, traders, and artisans could not always coincide. Groups that lived together were now too large to hold — if they ever had — their lands, tools or other resources in common. Some fields were more fertile or closer to the water sources than others, some craftsmen were better able to gain access to good materials, and some merchants more successful than their competitors. A new kind of pottery, symptomatic of the changing economic relationships, appeared in many Mesopotamian settlements. These coarse, ugly bowls with bevelled rims were manufactured in vast quantities, using a rudimentary means of mass production. Instead of being formed individually on a wheel, they were made of crudely worked clay that had been moulded to a uniform shape and size. For some time archaeologists thought that these vessels might have been used as containers for standardized rations of barley, oil or beer doled out to gangs of labourers as their wages; however, recent studies have demonstrated that although the bowls all seem to be of the same size, there are marked variations between them when it comes to the volume of grain or liquid they can hold.

Nevertheless, even though the bowls themselves can no longer be recruited as evidence to support the theory, this could well have been the point when society began to divide into employers and employees, into those who owned land and those who worked it. The basic building block of economic and social life may well have been the nuclear family, but this small unit was probably part of a clan — an extended family group who would combine their efforts to work an ancestral tract of land, or who specialized in the skills of a particular craft. Even within this kinship network there were probably certain individuals who would have wielded more power, and enjoyed greater authority, than others.

Inequalities of wealth and rank emerged, and could be seen nowhere more clearly than in the graveyards. The tombs of a rich minority were now resplendent with beautifully crafted metalwork, and with beads of ivory and lapis lazuli; the poor, in death as in life, made do with less. But even the lowliest Mesopotamian sought to provide comforts for loved ones in the darkness: a sparsely furnished child's grave near Mosul contained a little flute carved out of bone.

In archaeological terms, an ancient site qualifies for the term "city" rather than "town" when it shows evidence of having possessed the following: a population of thousands rather than hundreds; a central administration governed by king, nobles or priests, and housed in temples or palaces; a wide range of commercial and manufacturing enterprises and a wall around the perimeter to keep out enemies. The first human settlement which is known to have met all these criteria was the populous city of Uruk, lying on a branch of the Euphrates River some 250 kilometres southeast of modern-day Baghdad. The world had known large settlements before. Çatal Hüyük and Neolithic Jericho were not necessarily unique, but they left behind them only the faintest of traces on the landscape for later ages to find. When they vanished, so their way of life disappeared with them. But the inhabitants of Uruk left abundant traces of their lives and labours, and their city gave its name to an important stage in the development of Near Eastern civilization.

Uruk – the Biblical Erech – rose to prominence in the second half of the fourth millennium BC, when two adjacent communities, Eanna and Kullaba, expanded and coalesced. Here, between 40,000 and 50,000 people lived in close proximity in a settlement covering an area of 450 hectares, which would eventually be enclosed within a 10-kilometre city wall made of rugged, sun-scorched brick.

Uruk, in common with other Mesopotamian centres, acted as a powerful social magnet, and its population swelled at the expense of the smaller settlements in its rural hinterland. Communities that would once have functioned as entirely self-contained units now became little more than tributaries, sending their produce to the city, subordinate to the needs of the larger centre. Settlements that had previously flourished in the surrounding plain now declined in population, some even being deserted altogether. At the beginning of the fourth millennium BC there were at least 146 villages dotted around the countryside surrounding Uruk, each of which would have possessed its own temple, irrigation system and traditional clan-oriented social structure. Within 600 years that figure had shrunk to just 24, as generations of farmers and their families gradually drifted away to set up home in the city. There they could take up new careers – metalworking, sculpting, brickmaking – that would open up for them a way of life which was no longer subject to the vagaries of the climate or the predations of hungry nomads who might be driven by jealousy or desperation to attack a prosperous agricultural community.

So began the idea, still prevalent among many communities to this day, that the streets of cities were somehow "paved with gold", and that country dwellers were correspondingly less cultured and less intelligent than their urban counterparts. The first tangible sign of this notion appears in the great Sumerian epic poem *Gilgamesh*,

ACROSS LAND AND SEA

Humans used ships to reach Australia around 50,000 BC, but concrete evidence of people travelling by any means other than on foot does not begin until around 3500 BC. At this time, boats such as those overleaf carried goods and animals along the waterways of the Near East. At about the same time, the wheel, inspired perhaps by the earlier use of logs as rollers, transformed humans' ability to transport heavy loads on land. The battle wagon above, depicted on the soundbox of a lyre from a Mesopotamian royal tomb of the third millennium BC, shows that wheels were also useful in war.

From a fourth-millennium-BC grave in Hungary, this terracotta cup shaped like a wagon shows that the wheel spread rapidly after its first use in Mesopotamia or Transcaucasia.

An engraved stone seal of 3500 to 3000 BC shows a Mesopotamian cargo boat. The captain stands in the centre; a kneeling sailor steers, and another holds a punt pole, suggesting this was a canal or shallow-river craft. The leaf-tipped prow and stern are bent and tied down, showing the boat is made of reeds.

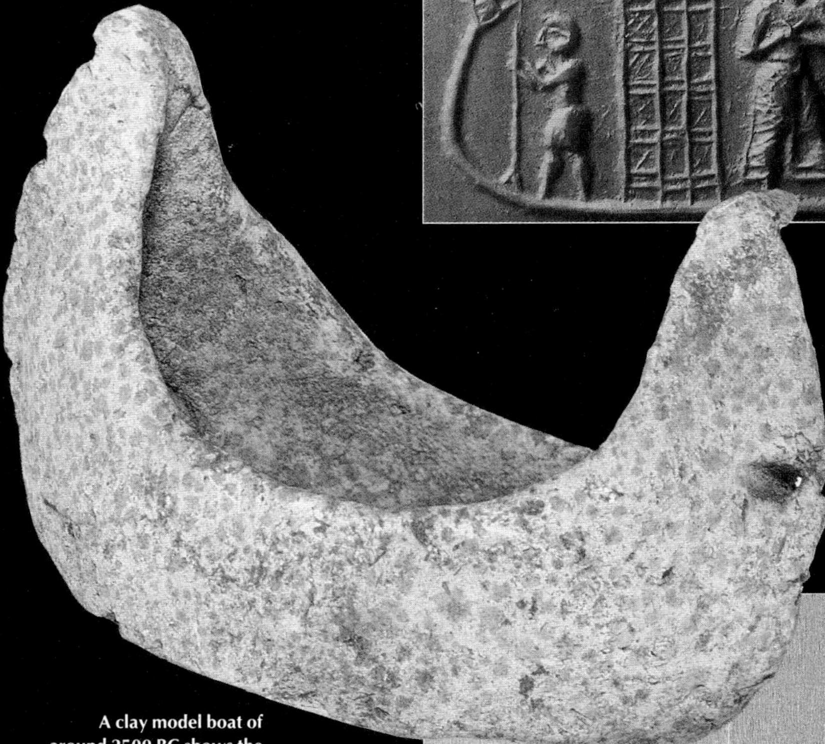

A clay model boat of around 2500 BC shows the high, curving prow and stern characteristic of Mesopotamian river craft.

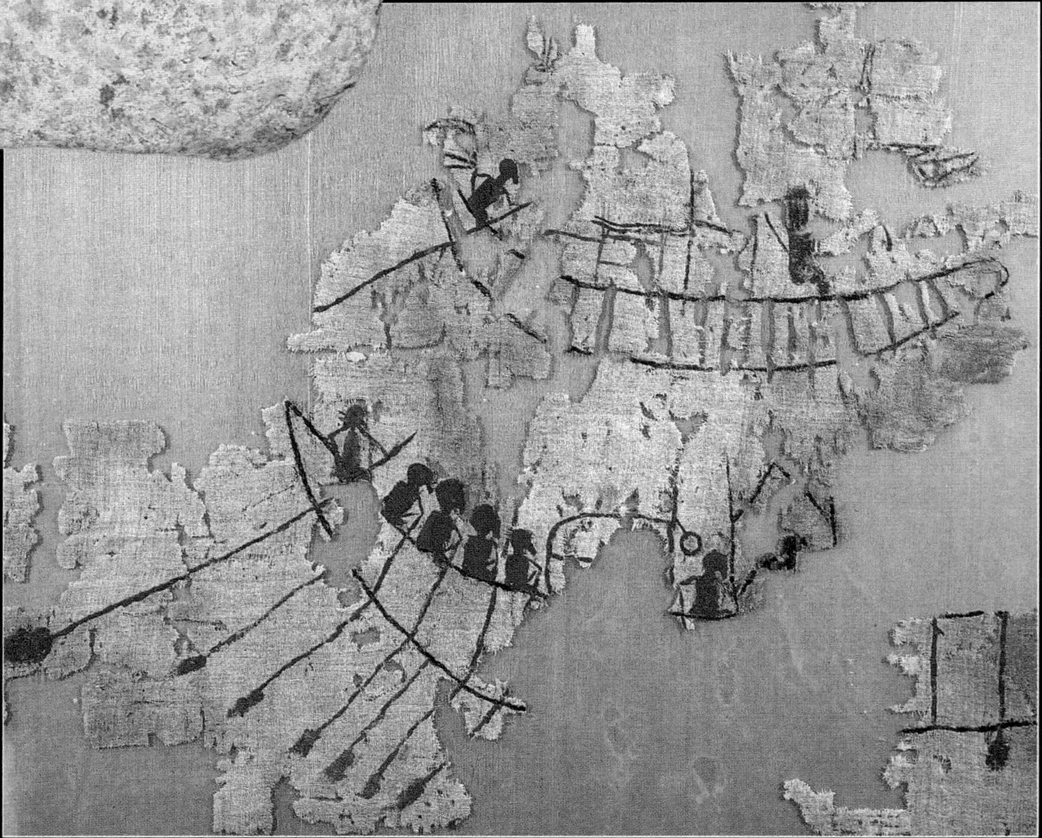

Fragments of linen found in an Egyptian cemetery are decorated with crescent-shaped Nile boats of the late fourth millennium BC. The long oars wielded by the bearded rowers in the lower boat suggest these were vessels designed to navigate deep rivers. The steering is provided by men at the stern, who are holding elongated paddles that serve as rudders. A central cabin offers shelter from the sun.

where a prostitute from the city taunts the unkempt Enkidu, who has been living in the mountains in preparation for his challenge to Uruk's overbearing king Gilgamesh. Scornfully the woman asks Enkidu why he chooses to run wild with animals out in the hills, when he could instead be in Uruk, where "all the people are dressed in their gorgeous robes, every day is a holiday and the young men and girls are wonderful to see. How sweet they smell!"

The lands around the city were owned by rich men who might have chosen to live either within the town or on their rural holdings, but the actual labour on the farms was carried out by peasants. These country estates had their own irrigation canals and date palm orchards, barns in which to house their draught animals and farming implements, and granaries used to store the annual harvest of wheat, barley, sesame and onions. They were large, well-organized units, possibly consolidating smaller parcels of land, and may have been held by clans or other forms of consortia, or by individual members of these groups.

Within the city itself there were ample signs of a community that was thriving. Uruk's most well-to-do inhabitants lived in large, two-storey houses with wooden balconies that ran the whole length of the upper floor. Here dwelt the prosperous merchants whose fleets of ships ploughed back and forth across the oceans to ports as distant as Bahrain and India. While out on the high seas their sailors were making money for them, the wealthy occupants of these grand houses could spend their evenings sitting out on their verandas, drinking beer brought to them by slave women and looking down over the rooftops of less prosperous neighbours. These were a mixture of bureaucrats, businessmen and professional craftsmen – maybe carpenters, scribes or weavers – who lived in single-storey houses that were less imposing than the adjoining mansions, but which nevertheless boasted a respectable number of rooms, all of which opened on to a secluded courtyard where convivial family parties might carry on late into the night.

Meanwhile, those who had less congenial homes, the poorer members of society such as fishermen, water carriers and unskilled building labourers, would find a comfortable spot where they could discuss the day's activities. Even the city's slaves – a mixture of foreign prisoners of war and impoverished rural Sumerians who had sold themselves for the sake of security – may have taken a sip or two from their masters' beer vats as they prepared the evening meal.

Carvings and statues give us some idea of what Uruk's citizens looked like. The men wore flowing curly beards and went bare chested; from the waist downwards hung knee-length, kilt-like garments. The women had their hair braided and oiled, and covered their bodies in generously cut gowns that concealed everything except their right arm, which remained naked to the air.

The city's prosperity depended on successful trade. Its merchants may have travelled far afield from Uruk in search of luxuries such as gold, electrum, ivory and lapis lazuli, but these were not merely adventurous shopping trips aimed at providing exotica that would please their rich clientele. A regular influx of new goods was crucial for the continued prosperity of the city as a whole. Uruk would not have been able to survive if the only raw materials available to its craftsmen and traders had been those immediately to hand in the plain around it, and if its customers had been limited to its closest neighbours.

Uruk's distinctive pottery was distributed so widely by trade that it could be found on sites all over Mescpotamia and beyond. So too could hundreds of ornately carved, cylindrical stone seals. These bore distinctive and carefully worked designs: plants, animals, strange amalgams of men and beasts, scenes of sacrifice, or even symbols of the gods themselves. The seals were pressed into specially prepared surfaces of clay, leaving their imprint. The clay impressions were then affixed, by cords, stiff reeds or strips of cloth, to bundles of merchandise. They served both as marks of ownership, and as address labels. Two different seal impressions – presumably indicating the identities of both vendor and purchaser – were often fastened together on a single parcel, to mark a bargain struck between the parties shown.

The appearance of these seals along the length of far-flung trade routes gave a reliable indication of the extent of Mesopotamian commercial activity. Most of the seals were cut from types of stone which could not be found naturally in Mesopotamia, and they turned up as far away as Egypt. It seems that even in the era before the Pharaohs ruled, there were already active contacts between the peoples that dwelt along the banks of the Euphrates and the Nile.

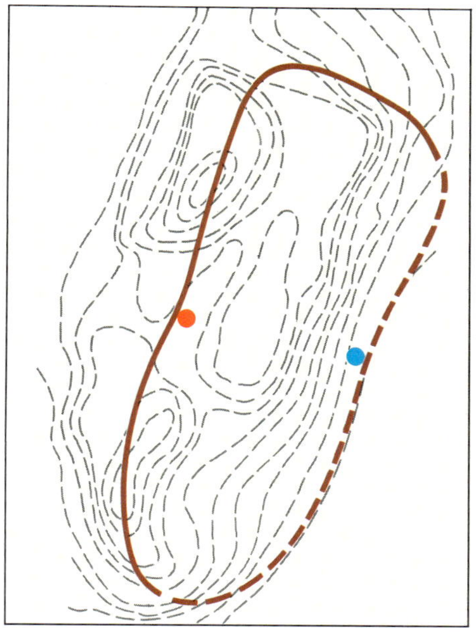

A massive stone watchtower guards the town of Jericho. Measuring at least 8.5 metres from top to bottom, the tower was built of rough-hewn stones bonded and faced with mud; access to the flat roof was gained via a steep stairway from an entrance at the base, as shown in the cutaway sections of this reconstruction. The plan above shows that the tower, indicated by the red dot, stood just within the town walls, as is thought to have been true of the abundant, year-round spring *(blue dot)*, although the part of the wall indicated by a thick, broken line has not been excavated. The dotted contour lines show the steep incline of the site today – the result of accumulated layers of occupation over several thousand years.

At the heart of Uruk, rising high above the bazaars and workshops, lay an impressive collection of ceremonial buildings. The first of these, the White Temple, occupied an ancient holy place. It stood on top of the ruins of a much older 'Ubaid shrine, an arrangement that provides powerful evidence for the existence of a deeply entrenched religious conservatism. New buildings might be erected, and the styles of their decoration might change, but the gods themselves and the rites that honoured them were old beyond the power of imagining.

At the White Temple a handsome triple staircase led up to the sanctuary itself, which was constructed to the same tripartite, cruciform plan that had been used in the region for centuries. Its imposing stepped altar faced a central table for the placement of burnt offerings. Additional staircases gave access to a flat roof, from which one could survey the whole city.

Four hundred metres to the east of the White Temple stood an even larger sacred area called the Eanna precinct. This was an extensive complex made up of broad terraces and courtyards, ceremonial halls, pillared porticoes and freestanding columns adorned with mosaics of brightly coloured terracotta, which glittered in the sunlight like serpents' scales. Hidden beneath these rich façades lay what served as the basic building blocks of Mesopotamia: rectangular mud bricks of even size and squared corners, with smooth surfaces which readily lent themselves to the application of elaborate forms of decoration.

The scale of these edifices was monumental. So, equally, was the obvious outpouring of wealth, skill, labour and administrative effort that was required first to construct and then to sustain them. As well as physically dominating the landscape by their sheer size, the shrines provided the focal point for every aspect of the city's varied life. Here, for the first time, was evidence of the existence of some form of centralized authority. While some Mesopotamians depended for their daily bread on secular employers and landowners, there were many others who owed their lives and their livelihoods to the gods; the temple ruled them.

The gods were the supreme sources of power, and the great temples their earthly residences. The priests who carried out ceremonial duties, and the administrators, scribes, stewards of farms belonging to the temple, and the craftsmen and traders at work on the temple's behalf, constituted a god's official household. In form and function, the houses of these divinities may have duplicated the domestic arrangements of ordinary mortals, although on a considerably grander scale. They were divided into private quarters, reception areas, storerooms, and offices where domestic and public business could be conducted.

The great temples of Mesopotamia, with their manifold ceremonial and economic functions, had evolved from the simple shrines of earlier ages to become gigantic ceremonial edifices. At Tell Brak, in northeastern Syria, there stood on top of the 40-metre mound that dominated the town an imposing temple dedicated to a cult in which the eye seems to have been the principal symbol. This impressive construction

must have been a source of wonderment to the inhabitants of the surrounding Habur plains. Visible from many kilometres away, the outer walls of the Eye Temple were covered in skilfully carved rosettes, the petals of which were made of white marble, black shale and pink limestone. Inside, the gleaming whitewashed walls were decorated with more rosettes, interspersed with red strips of limestone inlay and gleaming copper panelling. The inner sanctum housed an altar of coloured stone, bordered at the top and bottom by sheet gold which was fixed to the stone by gold-headed silver nails. Here worshippers would come in their thousands and leave small votive offerings of limestone idols, no more than six centimetres high and each with a pair — or pairs — of eyes carved into it.

But it was in the south that temple complexes were transformed so that they became virtual cities within a city. Uruk's Eanna precinct, for instance, not only included a temple but also housed living quarters for a sizable community of priests. In this theocratic society the priesthood formed a privileged and powerful elite. Its members controlled the religious mysteries, marshalled many of the community's resources, and monopolized specialized skills and knowledge that would be denied to laymen for centuries to come.

On lands owned by the temple, paid functionaries organized the collection of crops and the distribution of rations among the workers. Those who farmed the gods' fields handed over their produce in the expectation of earning not just earthly rewards but also divine goodwill. Sculpted reliefs show farmers entering the temple leading sacrificial beasts from their herds and flocks, and carrying baskets full to the brim with grain, dried fish and dates.

The building of the shrine must in itself have involved a massive public works project. With the experience they gained from overseeing the task, its priests probably went on to organize other collective efforts, such as the building of communal storehouses or the maintaining of irrigation canals. They probably would have sought to employ the town's leading craftsmen for the purpose, many of whom doubtless welcomed the opportunity to work in the service of the temple; apart from any spiritual favour temple employment might earn them, it would also provide them more immediately with a chance to work with the best and rarest materials. The ambitious design and construction of the huge sanctuaries would have both tested the mettle and advanced the techniques of bricklayers and builders, while the task of creating ritual objects of suitable grandeur clearly allowed bronze and goldsmiths to master all the metallurgical processes known to the age.

The temple's central position in the economic life of a large and vigorous population also furthered the development of systems for keeping records and calculating payments. To meet these diverse needs, a new specialist appeared — the *sanga*, a temple official whose title, in the Sumerian language of later generations, could mean either priest or accountant. To keep track of the goods that the faithful brought to the temple, the *sanga*s of Uruk attached clay seal impressions to offerings as a means of identifying their donors. They assessed the size of tithes or taxes, and organized the distribution of rations to the community at large. They had to know what was owed or promised, how much was given, what the goods were potentially worth in the context of any kind of barter or exchange, and what the temple had to pay out in return for any services rendered.

The priests who looked after these matters had to become adept in dealing with problems of quantity and value, and were likely pioneers in the development of

Sumerian mathematics, in which the number 60 was the basic unit of calculation. But even the most sophisticated numerical skills could not on their own suffice to keep the institution's affairs in order. The time had come to devise more efficient means of preserving and communicating complicated information.

For thousands of years, in all parts of the inhabited world, men and women had found ingenious ways of keeping track of simple facts. A shepherd might fill a pouch with pebbles – one stone for each sheep in the flock – or cut notches in a stick to count and remember the number of beasts in his charge. The earliest cave-dwellers had drawn pictures of humans and animals on their cavern walls. These images were the first steps in the gradual evolution of written signs and symbols; they were little more than the barest outlines, giving just enough visual information to impart a simple factual or ritual message.

In the same way that speech had evolved from the imitation of sounds, writing emerged from the graphic depiction of particular objects. The image that began as a rough but recognizable picture eventually turned into something more abstract: the memory of a memory of a more pictorial sketch. A simple system, based on a collection of descriptive marks, was apparently in use for thousands of years along the trade routes of western Asia. Clay tokens were found as far north as the Caspian Sea, as far south as Khartoum in Africa, and along an east-west axis that extended from

On the ivory handle of an Egyptian flint knife *(opposite page)*, factions clash in hand-to-hand combat with clubs and maces, while below them a naval battle rages. The knife dates from around 3400 BC, a time when the rulers of Upper Egypt coveted the more fertile lands of Lower Egypt. A thousand years later, war had become a sophisticated discipline, as shown in the images of Sumerian kings leading their troops into battle on the stone slab *(right)*. Mesopotamian city-states were in constant conflict: citizens were often conscripted to swell standing armies, and victories were consolidated by selling prisoners into slavery and installing governors in conquered provinces.

the Indus Valley to Anatolia. Over a period of 5,000 years or more the tokens changed very little, suggesting the existence of a widespread and enduring network of connections that embraced both the date palm farmers of southern Iraq and possibly even the original wall-builders of Jericho.

By the end of the fourth millennium BC, Mesopotamian priests were employing a rudimentary notation system that included many of the same ancient markings, which they incised on small clay tablets to be retained in the temple archives. As their communities expanded and transactions grew more complex, so too did the repertoire of signs necessary to keep track of them. In the form of protowriting that emerged in Uruk during this period, pictographs – recognizable images – were gradually joined by the more abstract symbols known as ideograms. By combining the two the temple scribes were able to represent general ideas in addition to numbers and concrete objects. The outline of a sun, for instance, not only signified the solar orb itself, but the related concept of day. The sign for arrow also functioned as the sign for life. As meanings became more elastic, the signs themselves slowly evolved into a cursive, linear script that was faster and easier to use, but contained less and less trace of its pictorial origins.

The system also had to be expanded so that it could accommodate some means of designating individual men and women. In a small, self-contained community, a simple mark, such as those used by potters to identify their work, would have been sufficient for the purpose. But the number of names in circulation in a city of several thousand souls required a more sophisticated solution, and signs that represented actual sounds of speech were gradually incorporated, thus allowing for the phonetic reproduction of different personal names.

The first applications of the art of writing were limited to financial purposes; only later were these early records of accounts and inventories supplemented by lists of gods. By 3000 BC trainee scribes learning their craft were having to memorize a collection of some 1,500 specific signs. These included important verbs such as to buy, and a range of nouns such as road, expedition, carpenter, smith, copper ingot, donkey, axe, harp, lyre and boat. There were also signs that could distinguish between two and four-wheeled chariots, and between two different kinds of plough. The vocabulary was not unique to the scribes of Uruk – the same lists, often written down in exactly the same sequence, were found in other contemporary Mesopotamian towns. Trade goods and similar pottery styles were no longer the only signs of the interactions that took place between far-flung communities. An articulate, intellectually active society was beginning to emerge.

For centuries this rudimentary writing would remain the exclusive preserve of the temples and their associated scribes. Their concerns inevitably dictated the content of the first written materials. But the rich legacy of decipherable texts that was left by the priests of Sumer would make it possible for later generations to acquire a more detailed understanding of their particular time and place than most other contemporary cultures were able to supply. And, belonging as it did to an era when social change came only very slowly, it was likely that the information recorded would have held true for many centuries, both before and after it was written down.

Until the arrival of these first written records, social hierarchies had been difficult to fathom. The men who appeared on carved reliefs, marching in ceremonial processions or performing ritual sacrifices, usually displayed no visible signs which indicated their rank. But even in these mute images, certain subtle social differences could still be detected. In the carved scenes of lion hunts and harvest festivals, one individual often seemed to be accorded more importance than the rest: he wielded the spear that killed the beast, was at the head of the parade of votive offerings, or otherwise stood apart from his companions.

No indication was given as to whether this putative hero was in fact a leader of the group or merely served as a symbolic figure. But it was clear that any authority he did possess was still of a religious rather than secular nature. He may have been the holder of the first title to emerge in early written texts, some time around 3000 BC — the *en*, or lord. The *en* apparently headed the temple establishment. He was just as much an administrator as a priest, being in charge of the community's economic activities and general welfare. He may well have performed ritual duties within the cult, for the texts in which the *en* appears suggest that among his various roles was the responsibility for ensuring that the forces of nature remained beneficent: if the harvest were blighted, or the flocks stricken by disease, the *en* was obliged to persuade the gods to put things right. Not all of the *en*s were male. In the city of Ur, at some point early in the third millennium BC, the title was held by a priestess.

The gods served by the *en* now began to stand out from one another as distinct individuals, with names, attributes and histories of their own. The Great Mother was still an object of reverence; the Mesopotamians called her Inanna, and worshipped her as the goddess of love, fertility, and ripening dates. Her emblem was the gatepost of a temple, and the pictograph of a gatepost eventually

Writing developed independently in America, China and Mesopotamia at different times, but appeared first of all in Mesopotamia during the fourth millennium BC. Its origins there lay in a system of accounting that used clay tokens such as those shown below. According to their shape and markings, the tokens indicated sheep, oil, perfume, wood or other commodities. They were dispatched with the goods in an envelope of clay *(left)*, and before being sealed inside were pressed against the soft clay, leaving indentations that the recipient could check against the tokens. From here, it was only a short step to marking on clay tablets with a stylus — and to extending the systems of markings so as to comprise a written vocabulary, as shown overleaf.

FIRST ACCOUNTS

became the written symbol of her name. The god Tammuz, who was depicted in later mythology as Inanna's spouse, represented the generative powers in nature: male animals rutting, or the springing seed. Other gods ruled over the sun and the moon, or presided over death, warfare, grain, flowing waters, flocks of sheep and herds of wild asses or gazelles, while a goddess who was particularly beloved by the Sumerians oversaw the fermentation of beer. Small settlements would have had their own cults, dedicated to the worship of obscure local deities.

At first, individual gods or goddesses were associated with particular cities. But as these communities expanded, and their temples established links between each other, the diverse histories and often overlapping identities of the gods became fused – perhaps with conscious intent on the part of a priestly elite – into a single mythological tradition. However, the belief that Sumerian cities were the property of specific gods persisted well into historic times. The temple did not necessarily own or control all the land in the district, but those who did so still owed their primary allegiance to the city and its deity – and, by logical extension, to the deity's representatives. The old tribal allegiances did not disappear, but became subsumed into a much larger relationship: that of an individual to his or her city. Yet there existed no word, in any of the ancient Mesopotamian languages that survived through written script, to express the concept of citizenship. Nor did any of these vocabularies acknowledge a distinction between village, town or city.

Mesopotamia had a large number of settlements that grew into cities. Uruk may not have been either the greatest or the wealthiest among them. According to later tradition, the city of Kish, 160 kilometres to the north, had been more important as an

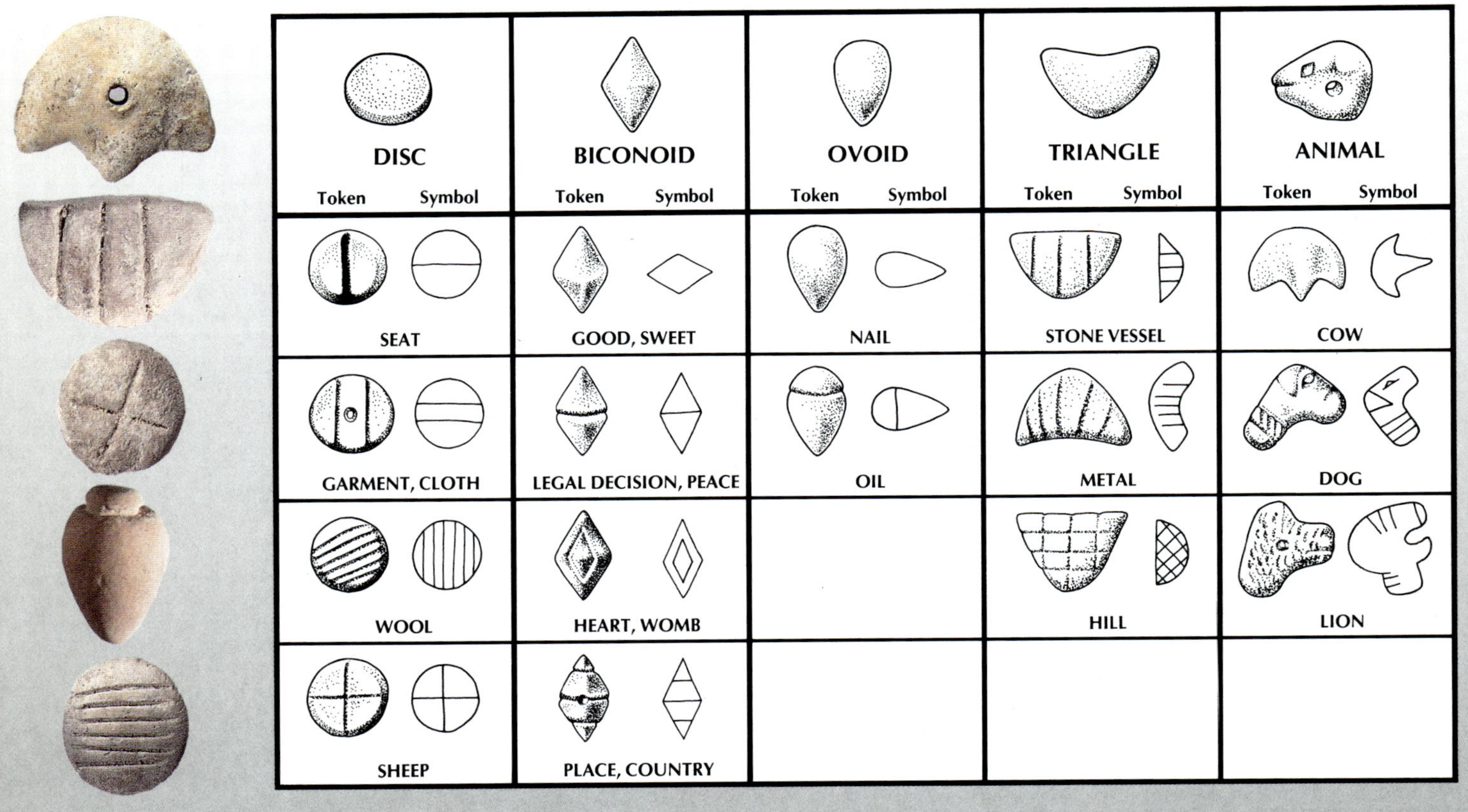

economic centre, while Nippur, 50 kilometres to the south of Kish and according to Sumerian mythology the home of Enlil, god of air, had greater religious significance.

Inhabitants of these large cities also founded colonies, such as Habuba Kabira in modern-day Syria, which lay some 900 kilometres north of southernmost Mesopotamia. Habuba Kabira occupied a valuable strategic site, controlling not only the trade routes to the south but also those leading northwest towards the Taurus Mountains and southern Anatolia and due west to the Mediterranean coast, just 150 kilometres away. Excavations at Habuba Kabira show every sign of a settlement that had been founded with a specific view to dominating the surrounding countryside. A massive fortifying wall, measuring some 1,000 metres long and three metres thick, with guard towers positioned every 30 metres along its ramparts and a second, smaller wall built in front for good measure, provided excellent protection on three sides of the rectangular-shaped town. On the fourth stood the broad Euphrates River, Habuba Kabira's main trading artery, along which goods evidently flowed in sufficient quantities to support a considerable number of businesses and workshops. The civic authorities had enough money to finance the construction of a central administrative and religious complex, along with irrigated gardens, well-ordered streets, and a sophisticated drainage system whereby water was drained from houses via clay pipes and into a canal network.

The careful town planning, along with the distinctive, mud-brick building style of southern Mesopotamia, shows that Habuba Kabira was in all probability designed and constructed by settlers who had originally come from one particular, although as yet unidentified, city somewhere in the south. Most of the other towns and settlements in Mesopotamia did not follow such a uniform pattern of development; each

The Sumerian clay commodity tokens shown on the opposite page represented, from top to bottom: cow, stone vessel, sheep, oil and wool. The chart shows how a token of a particular shape could be marked to convey different meanings, and how the shapes of tokens were transformed into written symbols that could be incised on a flat tablet.

Two clay tablets from around 3200 BC illustrate stages in the development of writing. The one above is probably a list of goods, with both impressed markings and inscribed characters. Only the latter appear on the tablet on the right – their meaning has not been fully deciphered, but they clearly express a more complex message.

one evolved at its own pace, and for its own different reasons. But, over a considerable length of time, a similar social structure came to prevail in all of them. Sumerian inscriptions written very early in the third millennium BC gave some indication of what this society might have been like.

Temple records listed the names of individual landowners, those families due to donate or receive rations, merchants conducting long-distance trade, or laymen who were carrying out certain administrative responsibilities. The documents also indicated that women played an active role in Mesopotamian society – they owned property, headed households, engaged in foreign commerce, and took their turn holding some form of public office.

Slowly the balance of power began to shift away from the temple and towards more secular government; it is likely that temples gradually began to share power with groups of clan elders who gathered together in some form of local assembly. Early Sumerian myths contained descriptions of councils of the gods, which presumably bore a strong resemblance to the existing political institutions. The members of these divine assemblies passed judgments and carried out sentences, and would hold emergency meetings in times of crisis.

Gradually, individual leaders emerged. The lord of the temple, the *en*, was now joined by two other titleholders: *lugal* and *ensi*. The term *lugal* meant some kind of great man or master; the contexts in which it was used made it unclear whether the *lugal* was a general term for an influential, prestigious public figure, or an actual king. The *lugal* apparently acted as a leader of men, but may have combined both secular and religious functions, since he was involved in the administration of the temple and the performance of certain rites within the cult.

The title *ensi* sometimes appeared in the same context as *lugal*, and it was occasionally used in a way that suggested that *ensi* may have been an alternative title – applied in some other city – for a man of the same high rank. Other texts, however, used *ensi* to indicate a figure who, while still lordly, was subordinate to some higher master. By the middle of the third millennium BC, the *ensi* was the title given to a city governor who was appointed to his post by some higher authority – which might possibly have been the temple, the assembly of elders, or a *lugal* who had become a power in the land.

Whatever name he was hailed by, the secular leader now started to assert his place within the Mesopotamian social structure. The earliest of the Sumerian myths and legends sketch a society where individual men began to claim royal status, either because they had been heroes who founded a city, or had been nominated by the gods. Later Sumerian king-lists hark back to leaders of some unspecified prehistoric period, such as Enmerkar, calling him "the man who built Uruk", or refer to a tradition that justifies kingly status by divine appointment. A monarch named Etana, for instance, is cited as the founder of the dynasty that ruled the great third-millennium-BC city of Kish; he was assigned by the gods to seek out the "plant of birth", the source of hereditary royal power, as a means of introducing the institution of kingship to the human race. To counter any potential challenges to their power, the kings made much of their divine patrons or parentage, conferring upon themselves such titles as "the beloved son of the goddess Ninhursag".

Relationships between these emergent kings and the old convocations of city elders may have been precarious. Gilgamesh, the Mesopotamian hero of the world's earliest known epic, claimed that it was the gods who had appointed him king of

Uruk. Nevertheless, when he wanted to go to war against his enemies, he was still compelled to seek the approval of the city's elders. They refused to grant it, so Gilgamesh took matters into his own hands and replaced them with a convocation of younger men who gladly authorized him to make war.

The need for strong military leadership served to bring individual rulers such as Gilgamesh into the limelight. Even if a council of elders reigned supreme over a city, they would inevitably have had to appoint some individual to lead their men on the battlefield. The rise of the individual, secular king may thus have been a reflection of the increased – and sustained – hostility that existed between communities. Cities looked beyond their borders; they were no longer content to rule only over their rural hinterlands. And, as any region grew in prosperity, so its settlements promised increasingly rich pickings for an invader. The war leader who had previously been appointed by an assembly to deal with only a temporary emergency now held his command on a more permanent basis. If he succeeded in keeping a city's enemies at bay, or in increasing the territory under its control, he might found a dynasty, ensuring that his sons would continue to rule after him.

The carved reliefs, stone seals and funerary monuments of the era spoke of a new obsession with warfare. The older pictures of hunting trips or peaceful harvest festivals were now joined by bloody battle scenes. The grave goods of the age also reflected this new concern with military matters: men of evident importance were now buried not just with a whole arsenal of shining weapons but also with the bodies of their armed retainers, who were put to death on their master's demise. The metalworkers would find the soldier to be as loyal and lucrative a customer as the priest. In the same period, the written sign for slave – deriving from a word meaning a foreign country – entered the scribal lexicon, while a seal from Uruk showed images of bound captives being marched off following a battle. In such a climate it was no wonder that the inhabitants of Gilgamesh's city organized labour gangs to construct and maintain a 10-kilometre-long wall.

The first steps towards a truly urban society in the Near East – defended settlements, far-reaching trade networks, irrigation systems, dominant religious institutions, specialized crafts, and incipient social and political hierarchies – came earlier than their equivalents in other parts of the world. But the tower builders of Jericho, the devout artists of Çatal Hüyük, the prosperous 'Ubaid farmers and Uruk's priestly elite were by no means unique. Much farther to the east, in the valley of the great Indus River that flowed down from the Himalayas, and beyond those same mountains in the vast expanses of China, their contemporaries were meanwhile travelling along their own distinctive paths towards civilization.

It was in the Kachhi plain, between the bare hills of Baluchistan in modern-day Pakistan and the fertile Indus Valley in northwest India, that the earliest flowering of Indian civilization occurred. The first people to inhabit the Kachhi plain were nomads from the Baluchistan mountains, who would descend each winter in order to escape the merciless cold that yearly swept over the craggy heights. Throughout the months when snow was coating their homelands, they were able to graze their herds on the lush Kachhi pastures and grow cereal for themselves on land they had fertilized with silt provided by dammed-up streams. Gradually, more and more nomads began to stay on when summer came, rather than retreating back to the hills; better to be hot in a land of plenty, they reasoned, than to be cool in a land of shortage.

Like tombstones in a vast burial ground, the standing stones of Carnac stretch across the Breton countryside. Shown here is a part of the site in which rough granite pillars, some of them four metres high, fan out from east to west in 12 straight rows.

Over a period of 2,500 years beginning during the fifth millennium BC, the first farmers of Europe marked their presence on the land by constructing ritual stone monuments on a stupendous scale. At Carnac, in the northwest of France, some 3,000 stones remain standing out of an original total of 10,000. Some of them are arranged in egg-shaped rings, while others are set up in straight lines which stretch for upwards of several hundred metres. Their careful alignment is believed to have charted in some unknown way the courses of the sun and moon.

THE MEGALITHS OF CARNAC

A stylized female outline *(right)* guards the entrance passage of a burial chamber at Carnac. Below, spiral carvings, perhaps reflecting lunar or solar orbits, adorn a passageway at Gavrinis, off Brittany.

Between 8000 and 5000 BC people on the western fringes of the Indus Valley system built up prosperous and long-lasting farming settlements. By the end of the sixth millennium BC, they were living in mud-brick houses, cultivating wheat and barley, and tending their thoroughly domesticated sheep, goats and cattle. So organized had their societies become that the village of Mehrgarh, which was built on a terrace overlooking the Bolan River, had a collection of storehouses that probably acted as central distribution points for food, craft materials and manufactured goods.

Like their Near Eastern contemporaries, the peoples of the Indus Valley fringe learnt how to manage the floods of the river so as to keep their crops watered. The prosperity that resulted from this allowed them to exploit fully their geographical position at the centre of a number of converging trade routes.

Their communities were small, but they had an appetite for exotic luxuries. They wore turquoise beads that may have originated from as far north as Turkmenia, in central Asia, and conch shells that came from the Arabian Sea, some 500 kilometres away. Among their most prized possessions were pendants made of perforated lead, and a few beads formed out of copper, which must have been a rarity, imported from some remote and unknown source. They would not progress as far as developing their own metallurgical skills, nor make any significant use of copper tools, until the beginning of the fourth millennium BC.

They placed the most desirable of their trade goods in the graves of the dead, along with human figurines of unbaked clay; food offerings were stored in reed baskets, coated with a protective layer of bitumen that made them as watertight as clay. As yet, the craft of pottery was still unknown to them, but when the craftsmen of the region did begin to work in ceramics, early in the fifth millennium BC, their design and decorations would echo Mesopotamian styles so strongly that the resemblances may well have been more than coincidental.

The upland valleys and the flood plains below them seem to have engaged in a constant flow of commercial and other exchanges. Communities may not have been large, but neither were they isolated or provincial. A large number of settlements grew and prospered because of their beneficial location at logical stopping-places on longer-distance trade routes. Caravans of merchandise, stopping for a rest or to engage in barter, might only swell such a settlement's population for brief periods, leaving no traces of their stay more permanent than the marks of their tent posts in the ground. But they became frequent and regular visitors, as reliable a source of wealth as any farmer's crops. To make their journeys possible, the traders needed these strategically placed towns as much as the local inhabitants needed them. Other settlements might come and go, but the caravan sites prospered – merchants moving goods through the western Indus Valley in 2000 BC passed through towns that had thrived continuously for 1,500 years.

To the west of the Indus Valley, towns such as Mundigak, in present-day Afghanistan, demonstrated movements towards an urban revolution that bore many resemblances to developments in the Near East. The town, on a well-trodden trade route into the heart of central Asia, would have witnessed the passage of lapis lazuli from the Afghan hills down to the coast, on its long journey to the jewellers of Sumer. Mundigak was already ancient before the third millennium BC began. Its inhabitants lived in well-built houses with hearths at the centre of many rooms. They had originally used stone to make tools and weapons, carving leaf-shaped arrowheads, but by 3000 BC they were taking an interest in metallurgy, working in copper and – by mid-millennium – in bronze. They loved vivid decoration. Their wheel-thrown pottery vessels glowed with painted polychrome designs – naturalistic scenes of animals, foliage and swimming fish, or handsome black geometric motifs against a deep red ground. They carved flat stone seals into intricate shapes and patterns – zigzags, mazes, diagonal grids and interlacing curves – using these, presumably, to print decorative designs on their woven cloth.

By 2500 BC Mundigak had grown into a substantial city. Like the great urban centres of Mesopotamia it boasted a large temple complex which rose above the town, but this edifice had a secular rival – an imposing brick-built palace, flanked by a colonnade of rounded pillars. But Mundigak's long and increasing prosperity had excited envy from others. In response to real or imagined threats of invasion, the whole town was now protected by imposing walls and square watchtowers, solidly constructed of sun-dried brick.

As the region's population swelled and new settlements sprang up in all directions along the Indus River system, so the interactions between these communities grew ever stronger. Trade may not have been the only reason behind the spread of skills such as pottery or metalworking, and the eventual disappearance of local differences in decoration in favour of more uniform styles.

As the material achievements of the Indus Valley culture extended over an enormous geographical area, so too did signs of its spiritual beliefs. For hundreds of years there had been female figurines in changing materials and styles, from rough shapes of unbaked clay to elaborately coiffed terracotta figurines. Soon after 3000 BC they were joined by animal companions: bulls and buffaloes, with six-petalled flowers or peepul-leaf garlands growing up between their horns. These deities, or descendants with strikingly similar attributes, would be worshipped throughout the Indian subcontinent well into modern times. Their devotees might be overrun or absorbed by waves of invaders or more peaceable migrants, but the gods themselves survived as proof of the extraordinary breadth and durability of the spiritual seeds first planted in the flood plain of the Indus River system.

Those seeds were to bear their richest fruit in the form of the first great Indian civilization that would flourish in the Indus Valley throughout the second half of the third millennium BC. With its two great cities of Harappa and Mohenjo-Daro, its strict, rectilinear street plans and its use of uniform weights and bricks, this society owed more than a little to the tradition of well-ordered Indus Valley communities that stretched back more than 5,000 years.

Between 7000 and 5000 BC, small farming communities scrabbled for their living in isolated clusters dotted along China's great rivers. There were few, if any, signs of contact between the separate settlements. The northerners lived in semisubterranean huts made of wattle and daub, and they cultivated millet, which they stored in pits. In the south, a landscape pockmarked with caves provided a rich supply of natural dwellings. However, their current occupants had very little in common with their Palaeolithic ancestors – they still hunted and fished for much of their food, but they had also learnt the techniques of growing rice.

The lives of southerners and northerners alike, in territories which were constantly vulnerable to floods and other natural disasters, must have felt dangerously precarious. They became early adepts in scapulomancy, a fortune-telling technique based on the application of heat to tortoise shells or animal bones, and the subsequent study of the cracks this process caused. The pattern of fractures was read like a map of the future, to see what lay ahead. To interpret these mysteries more precisely, diviners carved various symbols on the bones or shells before subjecting them to the fire. Inscriptions of this kind, made in a settlement in Henan province some time around 6000 BC, may have been the earliest pictographs of all, precursors of the Chinese characters that would not come into use until the second millennium BC. If so, the fortune-tellers of Henan would have anticipated the achievements of the scribes of Mesopotamia by at least 2,000 years.

In the centuries between 4000 and 3000 BC these disparate communities began to coalesce. Their inhabitants were no longer simply residents of a single village or members of a clan, but part of a much larger sphere of cultural contacts. During the next millennium the Chinese, too, began to congregate together in large, well-defended towns, in a society of specialized craftsmen and strongly differentiated

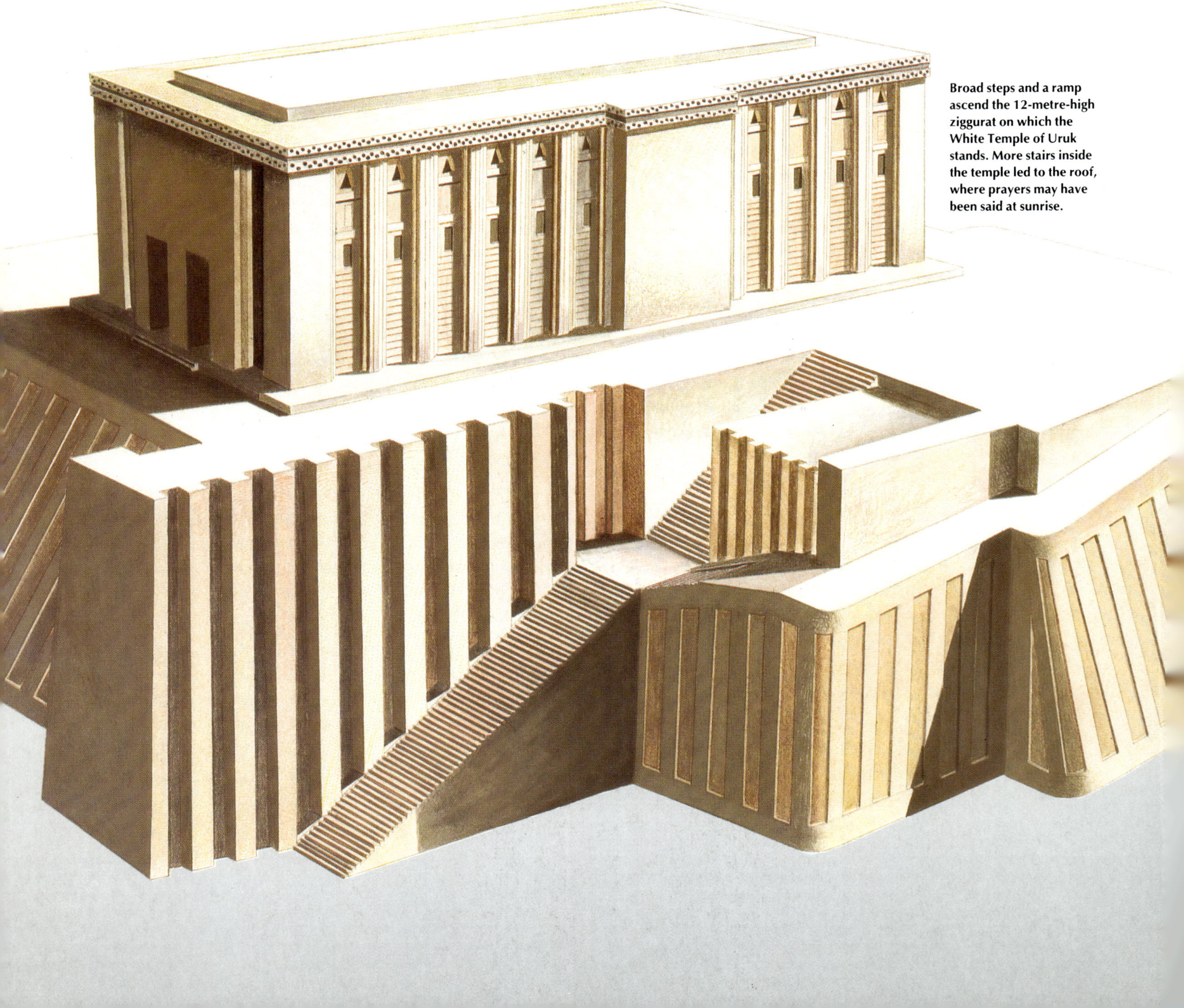

Broad steps and a ramp ascend the 12-metre-high ziggurat on which the White Temple of Uruk stands. More stairs inside the temple led to the roof, where prayers may have been said at sunrise.

social classes — a process of centralization that would eventually culminate in the establishment of China's first ruling dynasty, the Xia, soon after 2000 BC.

The inhabitants of Chengziyai, a populous community in Shandong province, lived in troubled times. They possessed an arsenal of bows and arrows that were clearly designed for warfare rather than simple hunting, and sought to protect themselves against some external threat by building a rampart of stamped earth, six metres high and nine metres wide. The creation of such a substantial structure suggested that the town had a large labour force at its disposal, yet it was unlikely that every inhabitant was equally obliged to dirty his or her hands: the graves in the town's cemetery gave evidence of marked differences of wealth and status.

Most of the burials were simple, rather spartan affairs, in which the corpse was placed, without being enclosed in any sort of coffin, in a narrow pit. The second largest group included in the pit a few simple grave goods. Nearby, however, were the tombs of their social superiors, whose bodies were buried in caskets, accompanied by jade ornaments and beautifully crafted pottery. These, however, were outshone in their turn by the graves of another, even more exclusive, group: the tombs of these most privileged citizens were capacious, fitted out with ledges and other architectural features, and lavishly endowed with luxury grave goods, including a curiously wrought, long-stemmed goblet, which was mounted on the carved jawbone of a pig.

Grave goods of the period suggested an elaborate system of religious beliefs. Ritual objects found in burials included oracle bones, like those used by the fortune-tellers of seventh-millennium Henan, and realistic bird and animal masks. Some well-furnished graves also contained secondary burials of additional skulls and limbs; these may have belonged to relatives of the original corpse, buried later, or to sacrificial victims who were dispatched to serve as attendants in some form of afterlife. Human sacrifice may also have been practised to the benefit of the living, particularly the members of the emergent upper

THE SEAT OF PRIESTLY POWER

Dominating the skyline of the early cities of Mesopotamia, massive temples proclaimed the high importance of the local deities — and also of the exclusive class of priests and priestesses who acted as the mediators between the people and their gods. The temple shown in the drawing on the opposite page was constructed sometime around 3200 BC at Uruk, a city that stood on the River Euphrates about 250 kilometres southeast of the site of modern Baghdad. The high platform, or ziggurat, on which it is built contained the bricked-in remains of earlier temples that had stood on the same site, which in this case was dedicated to the sky god Anu. Often lavishly decorated with elaborate clay mosaics coloured in black, red and white, these formidable constructions served to impress on the common people the political as well as the purely spiritual power of organized religion.

Processing around this alabaster vase from Uruk, naked priests bring the first fruits of the land to Inanna, the goddess of love and fertility. At the top of the vase she receives a jar brimming with produce.

classes; certain houses, better built and more substantial than their neighbours, contained infants buried in urns under the foundations or inside the walls.

The craftsmen who created goods for this prototypical aristocracy were sophisticated practioners of their arts. Potters could make eggshell-thin drinking cups, matching jars and lidded boxes, finished with a lustrous black glaze. Metalworkers had mastered the mysteries of copper — and, by the end of the third millennium BC, bronze. Stone-carvers had achieved equal heights in the treatment of jade. Varieties of jade were imported from widely scattered sources to achieve a range of visual effects; the stone was cut into fine rings, or carved into tubes — square on the outside, round within — which were incised with elaborate surface decorations.

The graves of the era gave evidence, too, of a less glorious development. In the village of Jiangou in Hubei province, at some point in the third millennium BC, dozens of corpses were dumped unceremoniously into a well. Their skulls were dented or shattered by heavy blows. Some bore the marks of posthumous scalping; other heads were missing altogether, apparently removed and carried off as trophies. The position of some remains suggested that a few people had been buried alive, and struggled vainly to climb out of the pit. Stone arrow and spearheads lay scattered around the site, near the ruins of a stamped earth wall. Brutal murder was nothing new; men had slaughtered each other since time began. But the carnage at Jiangou provided the first concrete evidence of mass violence in Chinese history.

In Hubei as in Jericho, on the banks of the Indus as on the banks of the Euphrates, men had become adept in yet another skill: the systematic waging of war. Some 7,000 years of progress on the part of the farmer, the smith, the trader and the temple priest had led, however indirectly, to this development. The farmer had learnt how to exploit and enhance the wealth of the land, making it valuable enough for others to covet. The bronze-worker had provided an arsenal of portable, efficient metal weapons, capable of far more devastating damage than cumbersome clubs or missiles of stone. The trader had beaten paths across unknown territory, establishing routes for invading armies to follow. And the priests of the temples had shown how to marshal large forces for collective enterprises, how to organize their labours, and how to distribute their rations and keep them at work.

Men would wage war for simple reasons: to claim new territory, to win control over water sources, to expand the power of their own group. The emerging social elites that dominated the great towns and their hinterlands now looked to extend their domains. Cities became city-states which themselves, in time, became nations and empires. Yet the greatest of them was not immune to threats from more powerful invaders. The proudest conquerors were inevitably conquered in their turn.

Civilization endured a bloody birth. The earliest written narratives recounted battles; the first individual names and faces to emerge into the light of history were those of military heroes and warrior kings. But ordinary men and women had greater and more peaceable victories to their credit. In the course of a few thousand years, they had gained some form of power over the world they lived in. Where they had once been nature's passive victims, they became her collaborators or, indeed, her conquerors. They had learnt how to farm, how to exploit the earth's mineral resources, how to make tools that would extend their physical capabilities, how to divide their labours to make fullest use of these new skills, and — most important of all — how to communicate and preserve their expanding knowledge. In so doing, they drew the blueprint for the modern world.

The enigmatic beauty of a young woman's face is captured in a white marble mask fashioned 5,000 years ago by an artist from Uruk in Mesopotamia. The skilful and sensitive modelling of the mouth, chin and cheeks, suggestive of a hitherto unknown interest in human character and physiognomy, was to remain unsurpassed in Mesopotamia for many centuries.

A Chronology of Major Events 5,000,000-30,000 BC

	5,000,000-3,000,000	3,000,000-2,000,000	2,000,000-1,000,000	1,000,000-500,000
EUROPE				First hominids reach Europe from Africa (850,000 BC).
THE AMERICAS				
AFRICA	Australopithecines, the earliest known bipedal hominids, inhabit the East African Rift Valley and southern Africa. First evidence of *Australopithecus afarensis* in eastern Ethiopia, dating from 3.5 million BC.	*Australopithecus africanus* inhabits southern Africa. First evidence for *Homo habilis*, who made use of stone tools, in eastern and southern Africa (2.3 million BC).	*Australopithecus robustus* inhabits southern Africa. Evidence of *Australopithecus boisei* at Olduvai Gorge in Tanzania. *Homo erectus* emerges in Africa around 1.6 million BC.	
NEAR EAST				Western Asia is inhabited by hominids.
FAR EAST				

500,000-200,000	200,000-100,000	100,000-50,000	50,000-40,000	40,000-30,000
Hunter-gatherers live in seasonal settlements such as that at Bilzingsleben in Germany. Evidence of human use of fire at Terra Amata, southwestern France.		*Homo sapiens neanderthalensis* (Neanderthal man) emerges in Europe.		*Homo sapiens sapiens* now present in Europe.
				Earliest evidence for human settlement at Pedra Furada in Brazil.
		Earliest evidence for *Homo sapiens sapiens*, anatomically modern man, in southern Africa.		
		Homo sapiens neanderthalensis present in regions from the Levant to central Asia.		
aves at Zhoukoudian in China re occupied by humans. vidence of the use of fire for ooking.			Humans reach Australia by boat from Indonesia.	

	30,000-20,000	20,000-15,000	15,000-10,000	10,000-9000	9000-8000
EUROPE	Cave paintings and Venus figurines bear witness to new artistic skills.			In southeast Europe and western Asia, tools and ornaments are made from beaten copper.	
THE AMERICAS				Humans reach the southern tip of South America. Both North and South America inhabited by big game hunters.	
AFRICA			Grindstones are used in Upper Egypt and Nubia to make flour from wild grass seeds.		Saharan rock art, including paintings of animals on rocks and in shelters, begins.
NEAR EAST					Sheep domesticated in northern Mesopotamia.
FAR EAST			First evidence of pottery at the Fukui cave in Japan.		

A Chronology of Major Events 30,000-3,000 BC

8000-7000	7000-6000	6000-5000	5000-4000	4000-3000
	Agriculture develops in the Balkans and west and central Mediterranean regions. Sheep, goats and cereals are domesticated.	Agriculture spreads as far north as the Netherlands. Technique of smelting is developed to extract copper from ore.	First megalithic monuments in northwestern Europe.	The simple plough is used in northern and western Europe. Evidence of wheeled vehicles in eastern Europe. Agricultural settlement founded at Skara Brae in the Orkneys.
Crops are cultivated in Mexico.	Crops are cultivated in Peru.			Earliest evidence of domestication of maize in Tehuacan Valley, Mexico. Domestication of the guanaco, a llama-like animal.
		Domestication of native North African cattle in the Saharan region.		
ully domesticated wheat, barley nd pulses cultivated in Fertile rescent.	Wheat, barley and pulses cultivated from Anatolia to Pakistan. Domesticated animals include goats, pigs and cattle. First evidence of textiles. Obsidian is widely traded in the Middle East. Çatal Hüyük, one of the first settled towns, is founded in Anatolia.	Evidence of the earliest known irrigation system in Mesopotamia. Northern Mesopotamia and Iran become centres of trade in painted pottery.	First evidence of the human use of sail in Mesopotamia. Domestication of the horse on the Eurasian steppes.	The world's first urban civilization emerges in southern Mesopotamia. First evidence for the use of boats and wheels in the Near East. Kilns and turntables are used by Mesopotamian potters. In Mesopotamia, evidence of the earliest writing.
		Farming villages established in China. Millet is cultivated, and pigs and dogs are kept. Wet rice farming established at Hemudu near the east coast of China.		Millet is cultivated in Korea.

BIBLIOGRAPHY

BOOKS

Aitchison, Leslie, *A History of Metals, Vol. 1.* London: Macdonald & Evans Ltd., 1960.

Bacon, Edward, ed., *Vanished Civilizations: Forgotten Peoples of the Ancient World.* London: Thames and Hudson, 1963.

Bahn, Paul G., and Jean Vertut, *Images of the Ice Age.* New York: Windward, 1988.

Baltscheffsky, H., H. Jornval, and R. Ringler, eds., *Molecular Evolution of Life.* Cambridge: Cambridge University Press/Royal Swedish Academy of Sciences, 1986.

Barker, Graeme, *Prehistoric Farming in Europe.* Cambridge: Cambridge University Press, 1985.

Barrow, J.D., and F.J. Tipler, *The Anthropic Cosmological Principle.* Oxford: Oxford University Press, 1988.

Bass, George F., ed., *A History of Seafaring Based on Underwater Archaeology.* London: Thames and Hudson, 1972.

Bender, Barbara, *Farming in Prehistory.* London: John Baker, 1975.

Benton, Michael, *Dinosaurs: An A-Z Guide.* London: Kingfisher Books, 1988.

Bordaz, Jacques, *Tools of the Old and New Stone Age.* Newton Abbot, Devon: David & Charles, 1971.

Bordes, François, *The Old Stone Age.* Transl. by J.E. Anderson. London: Weidenfeld & Nicolson World University Library, 1968.

Breuil, Abbé H., *Four Hundred Centuries of Cave Art.* Transl. by Mary E. Boyle. Montignac, Dordogne: Fernand Windels, 1952.

Brodrick, A. Houghton, ed., *Animals in Archaeology.* London: Barrie & Jenkins, 1972.

Burl, Aubrey:
Megalithic Brittany: A Guide to over 350 Ancient Sites and Monuments. London: Thames and Hudson, 1985.
Rings of Stone: The Prehistoric Stone Circles of Britain and Ireland. London: Weidenfeld & Nicolson, 1979.

Chang, Kwang-chih, *The Archaeology of Ancient China.* New Haven & London: Yale University Press, 1986.

Charleston, Robert J., ed., *World Ceramics.* London: Paul Hamlyn, 1968.

Childe, V. Gordon, *Skara Brae: A Pictish Village in Orkney.* London: Kegan Paul, Trench, Trubner & Co., 1931.

Clarke, David, and Patrick Maguire, *Skara Brae: Northern Europe's Best Preserved Prehistoric Village.* Scotland: Historic Scotland, 1989.

Clutton-Brock, Juliet, *Domesticated Animals from Early Times.* London: Heinemann & British Museum (Natural History), 1981.

Cox, Barry, Dougal Dixon, Brian Gardiner, and R.J.G. Savage, *Illustrated Encyclopedia of Dinosaurs and Prehistoric Animals.* London: Macmillan, 1988.

Craig, Annabel, *The Usborne Book of Prehistoric Facts.* London: Usborne Publishing, 1986.

Davis, Simon J.M., *The Archaeology of Animals.* London: B.T. Batsford, 1987.

Dennell, Robin, *European Economic Prehistory.* London: Academic Press, 1983.

Deutsches Archäologisches Institut, *Denkmäler Antiker Architektur.* Berlin: Verlag Walter De Gruyter & Co., 1982.

Dixon, J.E., J.R. Cann, and Colin Renfrew, *"Obsidian and the Origins of Trade"*, from *Avenues to Antiquity* by Brian M. Fagan. San Francisco: W.H. Freeman and Co., 1976.

Edwards, I.E.S., C.J. Gadd, and N.G.L. Hammond, eds., *The Cambridge Ancient History, Vol. 1, part 1: Prolegomena and Prehistory.* Cambridge: Cambridge University Press, 1970.

Fagan, Brian M.:
The Great Journey: The Peopling of Ancient America. London: Thames and Hudson, 1987.
New Treasures of the Past. London: Quarto Publishing Plc., 1987.
Readings from Scientific American: Prehistoric Times. San Francisco: W.H. Freeman and Co., 1983.

Foley, Robert, *Another Unique Species: Patterns in Human Evolutionary Ecology.* Harlow, Essex: Longman Scientific & Technical, 1987.

Foley, Robert, ed., *Hominid Evolution and Community Ecology: Prehistoric Human Adaptation in Biological Perspective.* London: Academic Press, 1984.

Frison, George C., ed., *The Casper Site: A Hell Gap Bison Kill on the High Plains.* New York: Academic Press, Inc., 1974.

Gamble, Clive, *The Palaeolithic Settlement of Europe.* Cambridge: Cambridge University Press (Cambridge World Archaeology), 1986.

Gimbutas, Marija, *The Gods and Goddesses of Old Europe: 7000 to 3500 B.C.* London: Thames and Hudson, 1974.

Goudie, A.S., *Environmental Change.* Oxford: Clarendon Press, 1983.

Gowlett, John, *Ascent to Civilization.* London: Collins, 1984.

Graziosi, Paolo, *Palaeolithic Art.* London: Faber & Faber, 1960.

Greenwood, P.H., *A Living Fossil Fish: The Coelacanth.* London: British Museum (Natural History), 1988.

Hall, H.R., C. Leonard Woolley, et al., *Ur Excavations, Vol. I: Al 'Ubaid.* Oxford: Oxford University Press, 1927.

Halstead, L.B., *Hunting The Past: Fossils, Rocks, Tracks and Trails.* London: Hamish Hamilton, 1982.

Hamblin, Dora Jane, and the Editors of Time-Life Books, *The First Cities* (The Emergence of Man series). Amsterdam: Time-Life Books, 1973.
Harding, Dennis, *Prehistoric Europe.* Oxford: Elsevier-Phaidon (The Making of the Past series), 1978.
Hawkes, Jacquetta, *The Atlas of Early Man.* London: Macmillan, 1976.
Hay, Richard L., *Geology of The Olduvai Gorge.* Berkeley: University of California Press, 1976.
Heinrich, Ernst, *Die Tempel und Heiligtümer Im Alten Mesopotamien.* Berlin: Verlag Walter De Gruyter & Co., 1982.
Howell, F. Clark, and the Editors of Time-Life Books, *Early Man.* Hong Kong: Time-Life Books (Life Nature Library series), 1984.
Howells, William W., "Homo Erectus", from *Avenues to Antiquity* by Brian M. Fagan. San Francisco: W.H. Freeman and Co., 1976.
Huyghe, René, ed., *Larousse Encyclopedia of Prehistoric and Ancient Art.* London: Paul Hamlyn, 1962.
Imbrie, John, and Katherine Palmer Imbrie, *Ice Ages: Solving the Mystery.* London: The Macmillan Press Ltd., 1979.
Institute of Vertebrate Palaeontology and Paleoanthropology, Chinese Academy of Sciences, *Atlas of Primitive Man in China.* Beijing: Science Press, 1980.
Jelínek, J., *The Pictorial Encyclopedia of The Evolution of Man.* London: Hamlyn, 1975.
Johanson, Donald C., and A. Edey Maitland, *Lucy: The Beginnings of Humankind.* London: Granada, 1981.
Kenyon, Kathleen M., *Excavations at Jericho. Vol 3: The Architecture and Stratigraphy of the Tell.* London: British School of Archaeology In Jerusalem, 1981.
Klein, Richard, "Ice-Age Hunters of Ukraine", from *Avenues to Antiquity* by Brian M. Fagan. San Francisco: W.H. Freeman and Co., 1976.
Kulke, Hermann, and Dietmar Rothermund, *A History of India.* London & Sydney: Croom Helm, 1986.
Kurten, Bjorn, *Pleistocene Mammals of Europe.* London: Weidenfeld & Nicolson, 1968.
Lamb, H.H., *Climate, History and the Modern World.* London: Methuen, 1982.
Lan-po, Chia, *The Cave Home of Peking Man.* Peking: Foreign Languages Press, 1975.
Leakey, Mary, *Disclosing the Past.* London: Weidenfeld & Nicolson, 1984.
Leakey, Richard:
The Making of Mankind. London: Michael Joseph, 1981.
One Life: An Autobiography. London: Michael Joseph, 1983.
Leakey, Richard, and Roger Lewin, *People of the Lake. Man: His Origins, Nature and Future.* London: Penguin Books, 1978.
Leroi-Gourhan, André:
"The Evolution of Paleolithic Art", from *Avenues to Antiquity* by Brian M. Fagan. San Francisco: W.H. Freeman and Co., 1976.
Préhistoire de L'Art Occidental. Paris: Éditions d'Art Lucien Mazenod, 1965.
Lumley, Henry de, "A Paleolithic Camp at Nice", from *Avenues to Antiquity* by Brian M. Fagan. San Francisco: W.H. Freeman and Co., 1976.
Manley, John, *Atlas of Prehistoric Britain.* Oxford: Phaidon, 1989.
Marshack, Alexander, *The Roots of Civilization: The Cognitive Beginnings of Man's First Art, Symbol and Notation.* London: Weidenfeld & Nicolson, 1972.
Mellaart, James:
Çatal Hüyük: A Neolithic Town in Anatolia. London: Thames and Hudson, 1967.
"A Neolithic City in Turkey", from *Avenues to Antiquity* by Brian M. Fagan. San Francisco: W.H. Freeman and Co., 1976.
Moody, Richard, *Prehistoric World.* London: Hamlyn, 1980.
Oates, David and Joan, *The Rise of Civilization.* Oxford: Elsevier-Phaidon (The Making of the Past series), 1976.
Piggott, Stuart:
Ancient Europe from the Beginnings of Agriculture to Classical Antiquity. Edinburgh: University Press, 1965.
"The Beginnings of Wheeled Transport", from *Avenues to Antiquity* by Brian M. Fagan. San Francisco: W.H. Freeman and Co., 1976.
The Earliest Wheeled Transport: From the Atlantic Coast to the Caspian Sea. London: Thames and Hudson, 1983.
Piggott, Stuart, ed., *The Dawn of Civilization: The First World Survey of Human Cultures in Early Times.* London: Thames and Hudson, 1961.
Pinna, Giovanni:
The Dawn of Life. London: Orbis Publishing (The World of Nature series), 1972.
Prehistory. London: Burke Publishing Co. Ltd., 1985.
Reader, John:
Missing Links: The Hunt for Earliest Man. London: Collins, 1981.
The Rise of Life: The First 3.5 Billion Years. London: Collins, 1986.
Renfrew, Jane M., "The Archaeological Evidence for the Domestication of Plants: Methods and Problems", from *The Domestication of Plants and Animals* by P. Ucko and G.D. Dimbley. London: Duckworth, 1969.
Richards, G., *Human Evolution: an Introduction for the Behavioural Sciences.* London: Routledge & Kegan Paul, 1987.
Savage, R.J.G., *Mammal Evolution: An Illustrated Guide.* New York & Oxford: Facts On File Publications, 1986.

Sherratt, Andrew, ed., *The Cambridge Encyclopedia of Archaeology*. New York: Crown Publishers, Inc/Cambridge University Press, 1980.
Simons, E.L., *Primate Evolution: An Introduction to Man's Place in Nature*. London & New York: Macmillan, 1972.
Sofer, O., and C. Gamble, *The World at 18,000 BP*. London: Unwin Hyman, 1989.
Sutcliffe, Antony J., *On the Track of Ice Age Animals*. London: British Museum (Natural History), 1985.
The Times, *Past Worlds: The Times Atlas of Archaeology*. London: Times Books, 1988.
Weisz, Paul B., *The Science of Biology*. New York: McGraw-Hill Book Company, Inc., 1963.
Willcox, A.R., *The Rock Art of Africa*. London & Canberra: Croom Helm, 1984.
Wood, Michael, *The World Atlas of Archaeology*. London: Mitchell Beazley, 1985.
Zervos, Christian, *L'Art de l'Époque du Renne en France*. Paris: Éditions "Cahiers d'Art", 1959.

PERIODICALS

Ackerman, S., "European Prehistory Gets Even Older". *Science* (Washington D.C.), No. 246, 1989.
Bahn, Paul G.:
"Early Teething Troubles". *Nature* (London), No. 337, 1989.
"The Unacceptable Face of the West European Upper Palaeolithic". *Antiquity* (Cambridge), Vol. LII, No. 206, 1978.
Benveniste, R.E., and G.J. Todaro, "Evolution of Type C Viral Genes: Evidence for an Asian Origin of Man". *Nature* (London), No. 261, 1976.
Bogucki, Peter, "The Antiquity of Dairying in Temperate Europe". *Expedition* (University of Pennsylvania), Vol. 28, No. 2, 1986.
Cann, R.L., et al., "Mitochondrial DNA and Human Evolution". *Nature* (London), No. 325, 1987.
COHMAP Members, "Climate Changes of the Last 18,000 Years: Observations and Model Simultations". *Science*, No. 241, August 1988.
Diamond, J.M., "Were Neanderthals the First Humans to Bury Their Dead?" *Nature* (London), No. 340, 1989.
Foley, Robert, and Robin Dunbar, "Beyond The Bones of Contention". *New Scientist* (London), October 1989.
Fricke, Hans, "Coelacanths: The Fish That Time Forgot". *National Geographic* (Washington, D.C.), Vol. 173, No. 6, June 1988.
Garrett, Wilbur E., "Where Did We Come From?" *National Geographic* (Washington, D.C.), Vol. 174, No. 4, October 1988.
Gladkih, Mikhail I., Ninelj L. Kornietz, and Olga Soffer, "Mammoth Bone Dwellings on the Russian Plain". *Scientific American* (New York), No. 248, 1983.
Katz, Solomon H., and Mary M. Voight, "Bread and Beer". *Expedition* (University of Pennsylvania), Vol. 28, No. 2, 1986.
Lewin, Roger:
"How Did Humans Evolve Big Brains?" *Science* (Washington, D.C.), No. 216, 1982.
"Man the Scavenger". *Science* (London), No. 224, 1984.
"Were Lucy's Feet Made for Walking?" *Science* (Washington, D.C.), No. 220, 1983.
Marshack, Alexander:
"Exploring the Mind of Ice Age Man". *National Geographic* (Washington, D.C.), Vol. 147, No. 1, January 1975.
"An Ice Age Ancestor?" *National Geographic* (Washington, D.C.), Vol. 174, No. 4, October 1988.
Marshall, J.C., "The Descent of the Larynx?" *Nature* (London), No. 338, 1989.
Putman, John J., "The Search for Modern Humans". *National Geographic* (Washington, D.C.), Vol. 174, No. 4, October 1988.
Renfrew, Colin, "The Origins of Indo-European Languages". *Scientific American* (New York), No. 261, 1989.
Rigaud, Jean-Philippe, "Art Treasures from the Ice Age Lascaux Cave". *National Geographic* (Washington, D.C.), Vol. 174, No. 4, October 1988.
Rukang, Wu, and Lin Shenglong, "Peking Man". *Scientific American* (New York), No. 248, 1983.
Simon, A.M., "The Last Neanderthal In France?" *Nature* (London), No. 287, 1980.
Straus, L.G., "Age of the Modern Europeans". *Nature* (London), No. 342, 1989.
Stringer, C.B., and P. Andrews, "Genetic and Fossil Evidence for the Origin of Modern Humans". *Science* (Washington, D.C.), No. 239, 1988.
Weaver, Kenneth F., "The Search for Our Ancestors". *National Geographic* (Washington, D.C.), Vol. 168, No.5, November 1985.

INDEX

Numerals in italics indicate an illustration of the subject mentioned.

References within entries to species of ape and early man are given in chronological, not alphabetical, order, eg Homo habilis, Homo erectus.

A

B

C

D

E

F

G

H

I

J

K

L

M

N

O

P

Q

R

S

T

U

W

Z

PICTURE CREDITS

The sources for the illustrations that appear in this book are listed below. Credits from left to right are separated by semicolons; from top to bottom they are separated by dashes:

Cover: Detail from a rock painting from Jabbaren in the region of Tassili-n-Ajjer, North Africa, P. Colombel/Photothèque du Musée de l'Homme, Paris. **2, 3:** Background artwork Roger Stewart/Ribbon artwork Oxford Illustrators. **9:** Giovanni Pinna, Milan. **10:** Map by Alan Hollingbery – Art by Colonel Mustard Design, Plymouth; Giovanni Pinna, Milan; Luciano Spezia, Milan. **11:** Giovanni Pinna, Milan (2). **12:** Luciano Spezia, Milan – Giovanni Pinna, Milan (2). **13:** Map by Alan Hollingbery – Prof. Dr. Bernard Krebs, Institut für Paläontologie, Freie Universität, Berlin; By courtesy of the Natural History Museum, London – Art by Colonel Mustard Design, Plymouth, Courtesy of Farish A. Jenkins Jr., Harvard University, Cambridge, Massachusetts – Art by Colonel Mustard Design, Plymouth, from a drawing by Prof. Dr. Zdeněk V. Špinar and Zdeněk Burian, Prague. **14:** Chip Clark, Smithsonian Institution, Washington DC. **15:** Map by Alan Hollingbery – Museo di Geologia e Paleontologia, Sezione del Museo di Storia Naturale, Universitá di Firenze, Italy: Art by Colonel Mustard Design, Plymouth. **16:** John Reader/Science Photo Library, London. **19:** Art by Bill Gregory. **20-21:** Art by Bill Gregory. **22-23:** Art by Town Group Consultancy, Figures by David Bergen. **24:** Prof. Dr. Hans Fricke, Max-Planck-Institut für Verhaltensphysiologie, Seewiesen; © Stephen Dalton/NHPA, London. **25:** Jack Dermid/Bruce Coleman Ltd., London; J.-P. Ferrero/Ardea London Ltd.; M.P.L. Fogden/Bruce Coleman Ltd., London. **26:** Art by David Bergen. **30:** John Reader/Science Photo Library, London. **34, 35:** Landscape Art by Jonothan Potter – John Reader/Science Photo Library, London; Art by David Bergen; Trustees of the British Museum, London; © Michael Holford/Collection British Museum, London. **37:** Art by David Bergen. **38, 39:** B. & C. Alexander, Dorset; Map by Town Group Consultancy, London. **40, 41:** Art by Martin Knowerden; Stephen J. Kraseman/Bruce Coleman Ltd., London (2); Art by Martin Knowerden; © Francois Gohier/Ardea London Ltd. **42, 43:** Art by Martin Knowerden; Sarah Errington/The Hutchison Library, London (background); Clem Haagner/Ardea London Ltd.; WWF/Timm Rautert/Bruce Coleman Ltd., London – Keith & Liz Laidler/Ardea London Ltd. **44, 45:** Art by Martin Knowerden; V. Southwell/The Hutchison Library, London (background); K.W. Fink/Ardea London Ltd. – Clem Haagner/Ardea London Ltd. – © Steven Kaufman/Bruce Coleman Ltd., London – Eric Dragesco/Ardea London Ltd. **46, 47:** Art by Martin Knowerden; Richard Woldendorp/Susan Griggs Agency, London (backgound); C.& S. Pollitt/NHPA, London; © Jean-Paul Ferrero 1980/Ardea London Ltd.; Art by Martin Knowerden; Hans & Judy Beste, Australia/Ardea London Ltd. **48:** Réunion des Musées Nationaux, Paris. **50:** Map by Alan Hollingbery. **52:** Art by Jon Blake **53:** Art by Colonel Mustard Design, Plymouth. **56, 57:** Art by David Bergen (2). **58:** Art by Max Schindler. **60-61:** Map by Alan Hollingbery. **63:** Jean Vertut, Paris. **64:** Art by Colonel Mustard Design, Plymouth. **65:** Trustees of the British Museum. **68-69:** Art by Bill Gregory (2). **73:** Réunion des Musées Nationaux, Paris (2). **75:** Photothèque du Musée de l'Homme, Paris. **78:** Gianni Dagli Orti, Paris (2). **79:** Erich Lessing/Magnum, London; © Ira Block Photography, New York 1989. **81:** Jean Vertut, Paris. **82-83:** Colorphoto Hans Hinz, Allschwil-Basel. **84, 85:** Jean Vertut, Paris (3). **86, 87:** Jean Vertut, Paris (3). **88, 89:** Jean Vertut, Paris (4). **90, 91:** Jean Vertut, Paris (4). **92:** Erich Lessing/Koszta József Museum, Szentes, Hungary. **94:** Maps by Alan Hollingbery. **96-97:** Maps by Alan Hollingbery. **98:** Art by Jon Blake. **101:** Aldo Durazzi, Rome/Courtesy Iraq Museum, Baghdad. **103:** The University Museum, University of Pennsylvania (Negs #T4 – 222, #T4 – 224c2) (2). **107:** Courtesy of the Oriental Institute of the University of Chicago, Illinois. **108:** Tohoku University from Heibonsha, Japan; Art by Colonel Mustard Design, Plymouth. **109:** Bulloz, Paris. **110:** Réunion des Musées Nationaux, Paris. **112:** Charles Tait, Orkney – National Museums of Scotland, Edinburgh. **113:** Map by Oxford Illustrators after a Crown Copyright original, published in V.G. Childe *Skara Brae* with permission – Charles Tait, Orkney; Historic Buildings and Monuments, Scotland. **114:** Photothèque du Musée de l'Homme, Paris. **116-117:** Foto Scala, Florence. **119:** © Christopher Chippindale, Cambridge. **121:** Arlette Mellaart/Ankara Museum of Anatolian Civilizations, Turkey. **122, 123:** Arlette Mellaart/Ankara Museum of Anatolian Civilizations, Turkey (3); Art by Tony Smith. **124, 125:** Art by Tony Smith; Arlette Mellaart/Ankara Museum of Anatolian Civilizations, Turkey (3). **126, 127:** Arlette Mellaart/Ankara Museum of Anatolian Civilizations, Turkey (2); Art by Tony Smith. **128:** British School of Archaeology in Jerusalem. **130-131:** Map by Alan Hollingbery. **132:** Trustees of the British Museum (3). **133:** Erich Lessing/Magnum, London. **134:** Hirmer Verlag, Munich. **138:** Trustees of the British Museum (2). **140-141:** Map by John Shipperbottom. **143:** © Michael Holford/Collection British Museum – Erich Lessing/Magnum, London. **144:** Iraq Museum Baghdad, Foto Staatliche Museen zu Berlin Vorderasiatisches Museum, Berlin – Trustees of the British Museum – Giacomo Lovera, Turin, Courtesy Museo Egizio, Turin. **146:** Art by Jonothan Potter. **147:** Map by Oxford Illustrators. **148, 149:** Réunion des Musées Nationaux, Paris (2). **151:** Denise Schmandt-Besserat, The University of Texas at Austin, Texas. **152:** Réunion des Musées Nationaux, Paris (5); Reprinted from Archaeology Magazine, vol. 32, no. 3. (© the Archaeological Institute of America). **153:** Staatliche Museen zu Berlin, Vorderasiatisches Museum, Berlin; C. M. Dixon, Canterbury. **156, 157:** Adam Woolfitt/Susan Griggs Agency (background); Corson Hirschfeld, Cincinnati, Ohio; C. M. Dixon, Canterbury (2). **160, 161:** Art by Andrew Robinson; Erwin Böhm, Mainz **163:** Richard Ashworth/Robert Harding Picture Library, London/Courtesy Iraq Museum, Baghdad.

ACKNOWLEDGEMENTS

The Editors wish to thank the following individuals and institutions for their valuable assistance in the preparation of this volume:

England: Bristol – Bob Savage, Professor of Vertebrate Palaeontology, University of Bristol. Cambridge – R.A. Foley, Lecturer in Biological Anthropology and Fellow of King's College; Jane M. Renfrew, Department of Archaeology, University of Cambridge; T.E.G. Reynolds, Evans Fellow, University of Cambridge; Sander van der Leeuw, Department of Archaeology, University of Cambridge. Hull – Paul Bahn. London – Roderick Conway Morris; Harriet Crawford, Institute of Archaeology, University College; Andy Current, The Natural History Museum; Liz Hodgson; Noah Klein; Ferdie McDonald; Caroline Manyon; James Mellaart, Institute of Archaeology, University College; Christine Noble. Southampton – Clive Gamble, Department of Archaeology, University of Southampton.

France: Paris – François Avril, Curateur, Département des Manuscrits, Bibliothèque Nationale; Hélène du Saussois, Départment des Antiquités Orientales, Musée du Louvre.

Hungary: Budapest – Pal Raczky, Department of Archaeology, Budapest University; Laszlo Selmetszy, Director, Törtely Museum, Budapest.

U.S.A.: Texas – Denise Schmandt-Besserat, Professor of Middle Eastern Studies, University of Texas at Austin. Wyoming – George C. Frison, Professor of Anthropology, University of Wyoming.

The index for this volume was prepared by Susanne Atkin.

Typeset by A.J. Latham Limited, Dunstable, Bedfordshire, England.
Printed and bound by Brepols S.A. – Turnhout, Belgium.
Colour separations by Fotolitomec, S.N.C., Milan, Italy.